Wild Vegetables: Morphology, Phytochemistry and Utility (Part 2)

Authored by

Ganesh Chandrakant Nikalje

*Department of Botany, Seva Sadan's R. K.
Talreja College of Arts, Science and Commerce
Affiliated to University of Mumbai, Ulhasnagar-3
India*

Apurva Chonde

*Department of Botany, Seva Sadan's R. K.
Talreja College of Arts, Science and Commerce
Affiliated to University of Mumbai, Ulhasnagar-3
India*

Sudhakar Srivastava

*Institute of Environment and Sustainable Development
Banaras Hindu University
Uttar Pradesh, India*

&

Penna Suprasanna

*Amity Centre for Nuclear Biotechnology
Amity Institute of Biotechnology
Amity University Maharashtra, Mumbai 410206
India*

Wild Vegetables: Morphology, Phytochemistry and Utility - *(Part 2)*

Authors: Ganesh Chandrakant Nikalje, Apurva Chonde, Sudhakar Srivastava & Penna Suprasanna

ISBN (Online): 979-8-89881-000-9

ISBN (Print): 979-8-89881-001-6

ISBN (Paperback): 979-8-89881-002-3

Published by Bentham Science Publishers Pte. Ltd. Singapore, in collaboration with Eureka Conferences, USA. All Rights Reserved.

First published in 2025.

need for a court order if at any point you breach any terms of this License Agreement. In no event will any delay or failure by Bentham Science Publishers in enforcing your compliance with this License Agreement constitute a waiver of any of its rights.

3. You acknowledge that you have read this License Agreement, and agree to be bound by its terms and conditions. To the extent that any other terms and conditions presented on any website of Bentham Science Publishers conflict with, or are inconsistent with, the terms and conditions set out in this License Agreement, you acknowledge that the terms and conditions set out in this License Agreement shall prevail.

Bentham Science Publishers Pte. Ltd.
No. 9 Raffles Place
Office No. 26-01
Singapore 048619
Singapore

Email: subscriptions@benthamscience.net

BENTHAM SCIENCE

CONTENTS

FOREWORD

Humans are dependent on plants for their food. Total 75% of the food supply to humans is drawn from just 12 crops and five livestock species. However, natural calamities, climate change, and other human activities pose a risk to the productivity of these species, with some potentially facing extinction. The ultimate goal of all scientists and policymakers is to see a hunger-free world. In this scenario, there is a need to expand the food base. Taking this into account, the book titled Wild Vegetables: Morphology, Phytochemistry and Utility by Dr. Ganesh Chandrakant Nikalje, Ms. Apurva Chonde, Dr. Sudhakar Srivastava, and Prof. Penna Suprasanna is a welcome step. In the global scene, there is a vogue to have plants as food from natural sources. I am happy to see the book with detailed information on the plants with their scientific name, names in different languages, their distribution, propagation and recipes. Many wild vegetables, especially leafy vegetables, have several essential elements like magnesium, calcium, sodium, etc. In villages and small towns like Anantapur, where I live, street vendors sell wild vegetables. The book gives detailed information on wild vegetables. The book also gives colour photographs for easy identification of wild vegetables. I am sure this book will be useful to both research scientists and laymen. This book will be a valuable resource for agriculturists and horticulturists to develop high-yielding varieties of these wild vegetables and for developing cultivation techniques. For nutritionists, it will be beneficial to fortify the human diet with vitamins and essential micronutrients.

I must congratulate all the four authors for this excellent book. I am sure this book will get a wider readership. This can be recommended to the students of Food Science and Nutrition.

Prof. T. Pullaiah
Sri Krishnadevaraya University
Andhra Pradesh, India

PREFACE

Biodiversity is an extremely important and balancing factor for the sustainable environment and ecosystems. Every single individual life form from the bacteria to higher evolved life forms is a component of Earth's ecosystem. Biodiversity refers to the relationship among various organisms and, encompasses diversity and ecosystem. Biodiversity constitutes the life support system for humankind for several needs including food, fodder, fuel, timber, pharmaceuticals and energy, and associated services (i.e. air and water, decomposition of wastes, recycling of carbon and nutrients, regulation of climate, regeneration of soil fertility, and maintenance of biodiversity).

Land use change, alterations in river flow, soil, water, and air contamination, misuse of marine resources, industrial activities, increasing population, and enhancing uniformity of food choices are the threats to biodiversity sustainability. The expansion of urban cities as well as rural agricultural activities has reduced the natural biodiversity-rich regions. The human's ability to adapt to environments and to interact with nature led to the use of wild plants and resources for their consumption in a sustainable manner. With the evolution of humans from the early hunter-gatherers to present times, and across unique variation ranges, plants have assumed extraordinary importance in human societies and, there is an interest in many wild species especially for meals and medicines. However, the increasingly prioritized choice for food sources (grains, fruits or vegetables) has promoted extensive cultivation of a few types of plants, while the rest of the plant types are becoming extinct or restricted.

Presently, nearly thirty domesticated crop species constitute a good-sized part of the dietary range and only three principal cereal grains (rice, wheat, and maize) make contributions to greater than half of the world's calorie consumption. While this is apparent, there are major cultivated vegetables, in which a number of 'fit to be eaten' species remained wild or semiwild, and had been left out during the process of domestication. However, these underutilized safe-to-eat elements have great potential to transform our food bowl in a more nutritious direction. The shift can enable the journey to a sustainable and climate change-resilient cultivation practice. The wild flora has played a very important role in contributing to the nutrition requirements of humankind all over the world and can continue to do so in the near future provided humans are aware of the potential accrued benefits.

Vegetables are consumed throughout the world for edible purposes. However, in the course of social and industrial evolution over the past few centuries, globalization has led to the homogenization of dietary habits. In the course of events, wild local relatives of a number of vegetables have been forgotten by the people and their consumption has decreased over the years. Such genetic resources of wild relatives of vegetables are decreasing and their cultivation is also getting reduced drastically. There is a need to impart knowledge to young students, researchers, and common people about the vast resources of wild relatives of vegetables in India.

This book focuses specifically on the Western Ghats, which is a huge reservoir of genetic reserve of a number of plants. The book provides ethnobotanical details, medicinal applications, phytochemical composition, and culinary notes of more than 120 wild vegetables belonging to 50 families. The information of wild vegetable plants is arranged alphabetically by family name, with each plant described in a consistent format. This book is divided in two volumes; the first volume consists of 23 families (Acanthaceae to Euphorbiaceae) and the second volume contains 27 families (Fabaceae to Zygophyllaceae).

The book shall act as a useful resource material for plant lovers, nature-enthusiasts, researchers and academicians, and those interested in food and nutrition.

Ganesh Chandrakant Nikalje
Department of Botany, Seva Sadan's R. K.
Talreja College of Arts, Science and Commerce
Affiliated to University of Mumbai, Ulhasnagar-3
India

Apurva Chonde
Department of Botany, Seva Sadan's R. K.
Talreja College of Arts, Science and Commerce
Affiliated to University of Mumbai, Ulhasnagar-3
India

Sudhakar Srivastava
Institute of Environment and Sustainable Development
Banaras Hindu University
Uttar Pradesh, India

&

Penna Suprasanna
Amity Centre for Nuclear Biotechnology
Amity Institute of Biotechnology
Amity University Maharashtra, Mumbai 410206
India

ACKNOWLEDGEMENTS

Authors are grateful to Himesh Jayasinghe, Jayant Kulkarni, Shrirang Bhutada, Anil Ingle, Ram Mane, Ajit Katkar, Vishal Patil, Rupal Wagh, Shivshankar Chapule, Rushikesh Khot, Shivaji Gade, Vikas Bhat, Suhas Kadu, Madhukar Bachulkar, Sunil Kadam, Pushpavati Bagul, Rakesh Sharma, Anil Phutane, Vasant Kale, Tukaram Kanoja, Vijesh Kumar, Rajesh Ghumatkar, Ganesh Pawar, Madhukar, Gaytri M. Chonde, Dnyanesh Kamkar, Sumit Bhosle, Suresh Shingare, Sumaiya Siddiqui, Madhura, Prajakta Nandgaonkar, Kailash Ugale, Pravin Ingle, Vishnu Birajdar, Trilok Barge, and Akshay Utekar for providing photographs of wild vegetables from their collection.

The authors are thankful to Shri. Dilip Shirodkar, Botanist and Shri. Shrikant Dadarao Ghodake, Senior Research Fellow, Flora of India, Botanical Survey of India, Western Regional Centre, Pune- 411001 for confirmation of plant photographs.

It is our pleasure to acknowledge, Dr. Ashwini Darshetkar, Post-Doctoral Fellow, Department of Botany, Savitribai Phule Pune University, Pune- 411007 for extending technical help in writing the book.

The authors are also thankful to their family members for their continuous support and encouragement in completing this book.

Ganesh Chandrakant Nikalje
Department of Botany, Seva Sadan's R. K.
Talreja College of Arts, Science and Commerce
Affiliated to University of Mumbai, Ulhasnagar-3
India

Apurva Chonde
Department of Botany, Seva Sadan's R. K.
Talreja College of Arts, Science and Commerce
Affiliated to University of Mumbai, Ulhasnagar-3
India

Sudhakar Srivastava
Institute of Environment and Sustainable Development
Banaras Hindu University, India

&

Penna Suprasanna
Amity Centre for Nuclear Biotechnology
Amity Institute of Biotechnology
Amity University Maharashtra, Mumbai 410206
India

INTRODUCTION

The United Nations Development Programme (UNDP) Sustainable Development Goals aim "to end poverty, protect the planet, and ensure that by 2030, all people enjoy peace and prosperity." However, extreme hunger and malnutrition continue to hinder progress in many parts of the world. In 2022, approximately 9.2% of the global population faced hunger, which increased from 7.9% in 2019, and about 2.4 billion people experienced moderate or severe food insecurity (FAO, IFAD, UNICEF, WFP, and WHO, 2023). Throughout history, humans have utilized a significant number of plant species, estimated between 40,000 and 100,000, for various purposes (IPGRI, 2002). Of these, around 30,000 are considered edible, and approximately 7,000 have been cultivated or collected for food (Asfaw, 2001; Arora, 2014). However, with the Green Revolution, many traditional crops were replaced by high-yielding varieties developed through breeding techniques, jeopardizing the diversity of plant species used for food and other purposes (Ebert, 2014; Guzo *et al.*, 2023). This loss in diversity may contribute to increased hidden hunger and undernourishment (Nikalje *et al.*, 2023). To circumvent such difficulties, diversifying food sources by increasing the usage of wild vegetables offers a promising strategy. Wild vegetables are naturally occurring plants suitable for human consumption, providing unique flavors and valuable nutrients distinct from cultivated plants. Wild vegetables thrive in diverse environments, such as forests, meadows, coastal areas, and deserts, and are often more resilient to harsh conditions, growing at minimal cost (Duguma, 2020). Despite their potential, wild vegetables remain underutilized and unavailable to the public at large, and are often limited to rural areas where they are most abundant (Leakey *et al.*, 2022). Increasing awareness and research on their domestication could promote sustainable agriculture, food security, and economic growth in rural communities (Luo *et al.*, 2022).

Wild vegetables can supplement human diets with proteins, essential minerals, and micronutrients, contributing to nutritional quality (Ogle, 2001). They provide an affordable source of nutrients for rural and semi-urban societies (Ickowitz *et al.*, 2016; Jones, 2017). Diverse diets are crucial for optimal nutrition, health, and well-being (FAO, WFP, and IFAD 2012). However, many low-income families in low- and middle-income countries consume staple-centric diets that lack diversity (Jones, 2017). Including wild edible foods in these diets could improve nutrition in an affordable way (Ickowitz *et al.*, 2016).

For indigenous and non-indigenous populations, wild edible plants serve as staple or complementary foods (Ju *et al.*, 2013). In rural regions, especially in drylands, they play a vital role in food security by filling seasonal gaps and serving as emergency foods during famines (Soromessa and Demissew, 2002). Many indigenous communities believe wild foods better maintain health. During periods of scarcity, over 70% of wild edible plants are consumed as stored food resources dwindle (Teklehaymanot *et al.*, 2010). Raising awareness about these plants could encourage their more frequent inclusion in diets, and support the rural economy.

However, several challenges limit their broader acceptance. The lack of knowledge about their identification, nutritional benefits, and safe preparation can deter people from consuming them. Limited seasonal availability, labour- intensive foraging, and the risk of mistaking edible plants for toxic look-alikes are additional barriers. Furthermore, some wild vegetables contain antinutritional compounds (*e.g.*, oxalates, tannins, and phytates) that can hinder nutrient absorption if not properly prepared (Ngurthankhumi *et al.*, 2024). Overharvesting can threaten their sustainability, and the absence of formal supply chains limits market availability. Finally, their strong or unfamiliar flavors may not align with consumer preferences, restricting their integration into modern diets.

The main intent of this book is to enhance efforts toward awareness and promote research on the domestication of wild vegetable plants. This could pave way for sustainable agriculture, food and nutritional security, and economic progression in rural areas.

REFERENCES

Asfaw, Z., Tadesse, M. (2001). Prospects for sustainable use and development of wild food plants in Ethiopia. *Econ. Bot,* 551, 47-62.
[http://dx.doi.org/10.1007/BF02864545]

Arora, R.K. (2014). *Diversity in underutilized plant species: an Asia-Pacific respective,* Biodiversity Internationalp. 203.

Duguma, H.T. (2020). Wild edible plant nutritional contribution and consumer perception in Ethiopia. *Int. J. Food Sci,* 2020, 1-16.
[http://dx.doi.org/10.1155/2020/2958623] [PMID: 32953878]

Ebert, A. (2014). Potential of underutilized traditional vegetables and legume crops to contribute to food and nutritional security, income and more sustainable production systems. *Sustainability (Basel),* 61, 319-335.
[http://dx.doi.org/10.3390/su6010319]

FAO, WFP, and IFAD (2012). The State of Food Insecurity in the World 2012 Economic Growth is Necessary but Not Sufficient to Accelerate Reduction of Hunger and Malnutrition. Rome: FAO.

FAO, IFAD, UNICEF, WFP and WHO. (2023). The state of food security and nutrition in the World 2023. Urbanization, agrifood systems transformation and healthy diets across the rural–urban continuum. (Rome: FAO).
[http://dx.doi.org/10.4060/cc3017en]

Guzo, S., Lulekal, E., Nemomissa, S. (2023). Ethnobotanical study of underutilized wild edible plants and threats to their long-term existence in Midakegn District, West Shewa Zone, Central Ethiopia. *J. Ethnobiol. Ethnomed,* 191, 30.
[http://dx.doi.org/10.1186/s13002-023-00601-8] [PMID: 37452368]

IPGRI. *Neglected and underutilized plant species: strategic action plan of the International Plant Genetic Resources Institute. ,* Rome, Italy.

Ickowitz, A., Rowland, D., Powell, B., Salim, M.A., Sunderland, T. (2016). Forests, Trees, and Micronutrient-Rich Food Consumption in Indonesia. *PLoS One,* 115, e0154139.
[http://dx.doi.org/10.1371/journal.pone.0154139] [PMID: 27186884]

Ju, Y., Zhuo, J., Liu, B., Long, C. (2013). Eating from the wild: diversity of wild edible plants used by Tibetans in Shangri-la region, Yunnan, China. *J. Ethnobiol. Ethnomed,* 91, 28.
[http://dx.doi.org/10.1186/1746-4269-9-28] [PMID: 23597086]

Jones, A.D. (2017). Critical review of the emerging research evidence on agricultural biodiversity, diet diversity, and nutritional status in low- and middle-income countries. *Nutr. Rev,* 7510, 769-782.
[http://dx.doi.org/10.1093/nutrit/nux040] [PMID: 29028270]

Luo, G., Najafi, J., Correia, P.M.P., Trinh, M.D.L., Chapman, E.A., Østerberg, J.T., Thomsen, H.C., Pedas, P.R., Larson, S., Gao, C., Poland, J., Knudsen, S., DeHaan, L., Palmgren, M. (2022). Accelerated domestication of new crops: Yield is key. *Plant Cell Physiol,* 6311, 1624-1640.
[http://dx.doi.org/10.1093/pcp/pcac065] [PMID: 35583202]

Leakey, R., Tientcheu Avana, M.L., Awazi, N., Assogbadjo, A., Mabhaudhi, T., Hendre, P., Degrande, A., Hlahla, S., Manda, L. (2022). The future of food: Domestication and commercialization of indigenous food crops in Africa over the third decade (2012–2021). *Sustainability (Basel),* 144, 2355.
[http://dx.doi.org/10.3390/su14042355]

Ngurthankhumi, R., Hazarika, T.K., Zothansiama, , Lalruatsangi, E. (2024). Nutritional composition and anti-nutritional properties of wild edible fruits of northeast India. *Journal of Agriculture and Food Research,* 16, 101221.
[http://dx.doi.org/10.1016/j.jafr.2024.101221]

Nikalje, G.C., Rajput, V.D., Ntatsi, G. (2023). Editorial: Putting wild vegetables to work for sustainable agriculture and food security. *Front. Plant Sci,* 14, 1268231.
[http://dx.doi.org/10.3389/fpls.2023.1268231] [PMID: 37841612]

Ogle, B.M., Grivetti, L.E. (1985). Legacy of the Chameleon: Edible wild plants in the Kingdom of Swanziland, part I-II. *Ecol. Food Nutr,* 17, 1-64.
[http://dx.doi.org/10.1080/03670244.1985.9990879]

Soromessa, T., Demissew, S. (2002). Some uses of plants by the Benna, Tsemay and Zeyise people, Southern Ethiopia. *Ethiopian Journal of Natural Resources,*

KEYWORDS

Wild vegetables, Identification characters, distribution, flowering/fruiting season, propagation, chemical constituents, recipe, uses, Dietary supplements, Alkaloids. Flavonoids, vitamins, minerals, saponins, steroids, terpenoids, Anti-inflammatory, Analgesic, Antimicrobial, Anti-diabetic, Antioxidant, Hepatoprotective, Anti-cancer, Anti-hyperlipidemic, wound healing, antipyretic, diuretic, stomachic, laxative, biliousness, leprosy, bronchitis, leucorrhea, hysteria, tonsillitis, malaria, dysentery, dysuria, chicken pox, fever, mania

CHAPTER 1

Wild Vegetables of the Family Fabaceae

INTRODUCTION

This family is often called the legume or bean family. It is estimated to contain around 20,000 species across over 700 genera. The members are a vital source of food for humans and animals worldwide due to its high nutritional value. They are a rich source of plant-based proteins, dietary fibers, carbohydrates, vitamins and minerals including iron, folate, potassium, and phosphorus, *etc.* Some members contain anti-nutritional compounds such as lectins, phytates, tannins, *etc.* (Martín-Cabrejas 2019).

Bauhinia malabarica Roxb.

Botanical name: *Bauhinia malabarica*

Family: Caesalpiniaceae

Local name: Koral, Korat

Vernacular name:

- **Assamese:** Kotora
- **Bengali:** Karmai
- **English:** Malabar bauhinia
- **Hindi:** Amli
- **Kannada:** Basavanapaada
- **Malayalam:** Arampuli
- **Oriya:** Gumbati
- **Sanskrit:** Amlapatrah
- **Tamil:** Puli-y-atti
- **Telugu:** Puli Chinta

Season: June & July

Parts used: younger stem and leaves

Ganesh Chandrakant Nikalje, Apurva Chonde, Sudhakar Srivastava & Penna Suprasanna

Characteristics:

1. Leaves: The leaves are alternate and compound. They are bilobed or bifid, which means the leaf blade is divided into two lobes that are usually rounded or heart-shaped. The lobes are joined near the base, giving the appearance of a butterfly's wings or a cow's hoof, which is a characteristic feature of the *Bauhinia* genus (Sharma *et al.*, 2014) (Fig. **24.1**).

Fig. (24.1). Leaves of *B. malabarica* (PC: Prajkta Nangaonkar).

2. Flowers: The flowers are large and showy, typically 3-5 cm in diameter. They are bisexual and have five petals that are often white, cream, or pale yellow in color. The petals are arranged in a shape resembling an open butterfly, with one petal larger and broader than the others (Sharma *et al.*, 2014).

3. Inflorescence: The flowers are arranged in terminal or axillary racemes or panicles. They form clusters of several flowers, which bloom sequentially along the inflorescence (Sharma *et al.*, 2014).

4. Fruits: The fruit is a legume or pod. It is elongated, flattened, and woody when mature. The pod contains several seeds and often splits open when ripe to release the seeds (Sharma *et al.*, 2014).

5. Bark and Branches: The bark is smooth and grayish-brown in color. The branches are slender and spread out in a zigzag pattern (Sharma *et al.*, 2014).

Distribution:

Indo-Malesia; Indian distribution- Assam, Madhya Pradesh, Meghalaya, Odisha. Maharashtra: Nasik, Pune, Raigad, Sindhudurg, Thane.

Propagation: Seed and Cuttings

Chemical constituents:

Seven flavonols, including 6,8-di-C-methyl kaempferol 3-methyl ether, kaempferol, afzelin, quercetin, isoquercitrin, quercitrin, and hyperoside were isolated from the methanol extract of leaves (Rawiwun *et al.*, 2008).

Recipe:

Ingredients: Moong dal, Korla leaves, mustard, cumin, green or red chilies, turmeric powder, hing, garlic.

Method: Cooked with soaked moong dal. Use only leaves; remove tough veins, mainly the central ones from mature leaves. Wash, and chop the leaves. Heat oil, add a small amount of mustard and cumin, then hing (or garlic, whichever is preferred), green/red chilies. Add the soaked dal (drain first) and then the leaves, turmeric, and salt to taste. Cook until it becomes tender.

Uses:

1. Traditional Medicine: Various parts of plants, including the bark, leaves, flowers, and roots, have been used in traditional medicine systems for their potential medicinal properties. It possesses antioxidant, anti-inflammatory, antimicrobial, analgesic, and anti-diabetic properties. The plant has been used to treat fever, wounds, skin infections, digestive disorders, respiratory ailments, diabetes, *etc.* (Thenmozhi *et al.*, 2013; Sharma *et al.*, 2014).

2. The young shoots of *B. malabarica* are edible and are commonly prescribed to treat cough, gout, glandular swellings and goiter, haemorrhage, leprosy, menorrhagia, scrofula, urinary disorders, wasting diseases, worm infestations and for liver disorders (Venkatachalapathi *et al.*, 2015).

3. Wound Healing: The bark is often used in traditional medicine for its potential wound-healing properties. It may be applied topically to wounds and injuries to promote healing and reduce inflammation (Ahmed *et al.*, 2012).

4. Ornamental Plant: It is also cultivated as an ornamental plant for its attractive flowers. The showy, butterfly-shaped flowers and the overall aesthetic appeal of

the tree make it a popular choice for landscaping and gardens (Venkatachalapathi *et al.*, 2015).

5. Timber: The wood is reported to be hard and durable. In some regions, it is used for construction purposes, making furniture, and crafting small wooden items (Venkatachalapathi *et al.*, 2015).

Bauhinia racemosa Lam.

Botanical name: *Bauhinia racemosa* Lam.

Family: Fabaceae

Local name: Apta, Sona

Vernacular name:

- **Hindi:** Kathmauli, Jhinjheri
- **Marathi:** Aapta, Sona
- **Tamil:** Atti, Tataki
- **Malayalam:** Arampaali, Kutabuli, Malayaththi
- **Telugu:** Tella arecettu
- **Kannada:** Aapta, Aralukadumandara
- **Bengali:** Banraji, Banraj
- **Konkani:** Apto
- **Sanskrit:** Yamalapatrakah, Yugmapatra
- **Urdu:** Gul-e anehnal

Season: Flowering: February-May

Parts used: All parts

Characteristics (Fatima *et al.*, 2021):

1. Leaves: The leaves are alternate and compound. They are bilobed or bifid, which means the leaf blade is divided into two lobes that are usually rounded or heart-shaped, resembling a butterfly's wings or a cow's hoof. The lobes are joined near the base, creating a characteristic "U" shape (Fig. **24.2**).

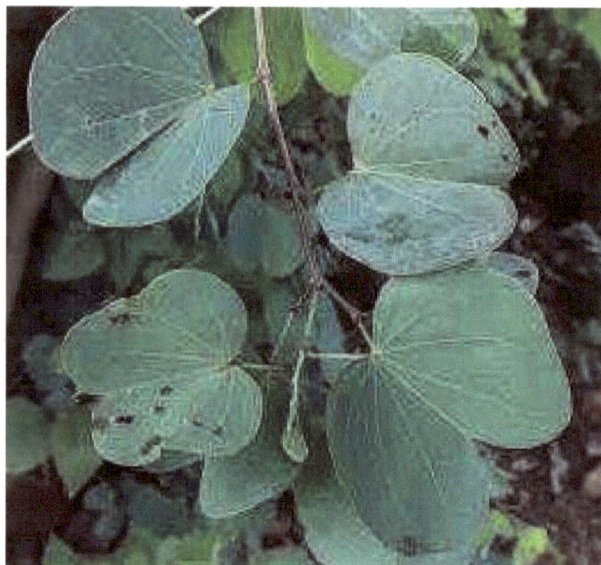

Fig. (24.2). Leaves of *B. racemosa* (PC: Apurva Shankar Chonde).

2. Flowers: The flowers are typically large and showy. They are bisexual and have five petals that are usually yellow or yellowish-orange in color. The petals are arranged in a shape resembling an open butterfly, with one petal larger and broader than the others.

3. Inflorescence: The flowers are arranged in racemes, which are elongated clusters of flowers along a central stem. The racemes can be terminal or axillary, meaning they can be found at the tips of branches or in the leaf axils.

4. Fruits: The fruit is a legume or pod. It is elongated, flattened, and woody when mature. The pod contains several seeds and often splits open when ripe to release the seeds.

Distribution:

E. Asia - China, India, Myanmar, Thailand, Cambodia, Vietnam.

Propagation: Seed, Stem Cuttings

Chemical constituents:

The major bioactive components of *B. racemosa* are methyl gallate, gallic acid, kaempferol, quercetin, quercetin 3–O–α–rhamnoside, kaempferol 3–O–β–glucoside, myricetin 3–O–β–glucoside and quercetin 3–O–rutinoside (Rashed *et al.*, 2014).

Recipe:

Ingredients: Moong dal, Korla leaves, mustard, cumin, green or red chilies, turmeric powder, hing, garlic.

Method: Cooked with soaked moong dal. Use only leaves; remove tough veins, mainly the central ones from mature leaves. Wash, and chop the leaves. Heat oil, add a small amount of mustard and cumin, then hing (or garlic, whichever is preferred), and green/red chilies. Add the soaked dal (drain first) and then the leaves, turmeric, and salt to taste. Cook until it becomes tender.

Uses:

1. Antioxidant Activity: Some studies have suggested that *B. racemosa* may possess antioxidant properties, which can help protect against oxidative stress and damage caused by free radicals in the body (Panda *et al.*, 2015).

2. Anti-inflammatory Potential: It has been investigated for its potential anti-inflammatory properties. It may exhibit inhibitory effects on certain inflammatory markers, which could be beneficial in managing inflammatory conditions (Panda *et al.*, 2015).

3. Antimicrobial Activity: Several studies have explored the antimicrobial properties of *B. racemosa*. Extracts from the plant have shown activity against certain bacteria, fungi, and parasites, indicating potential use in combating microbial infections (Panda *et al.*, 2015).

4. Antidiabetic Effects: It has been investigated for its potential antidiabetic effects. Some studies suggest that extracts from the plant may help regulate blood sugar levels and improve insulin sensitivity, which could be beneficial for individuals with diabetes.

5. The bark and leaves of *B. racemosa* are sweetish and acrid, used as a refrigerant, and astringent, in the treatment of headache, fever, skin diseases, blood diseases, dysentery, and diarrhea. A decoction of the bark is recommended as a useful wash for ulcers (Panda *et al.*, 2015).

6. Ayurveda: The plant is commonly used for the treatment of the initial stage of cancer (Wealth of India 1953; Kirtikar *et al.*, 1975).

Canavalia ensiformis (L.) DC.

Botanical name: *Canavalia ensiformis*

Family: Fabaceae

Local name: Abai cha shenga

Vernacular name:

- **Common name:** Horse Bean, Jack bean, Sword bean
- **Kannada:** Tamate balli
- **Marathi:** Abai
- **Mizo:** Fangra

Season: late spring to early autumn

Parts used: Seeds, young pods, leaves, stem

Characteristics (Acevedo-Rodríguez, 2005):

1. Plant Type: *C. ensiformis* is a fast-growing annual or perennial vine.

2. Leaves: The leaves are alternate and trifoliate, consisting of three leaflets. Each leaflet is elongated and lanceolate in shape, with a smooth or slightly hairy surface. The leaflets can measure around 7-15 centimeters in length.

3. Flowers: The flowers are pea-like and typically purplish or pinkish in color. They are borne on long, slender stalks and occur in clusters or racemes.

4. Pods: The mature pods of *C. ensiformis* are long, slender, and curved, resembling a sickle or sword. They can reach lengths of 15-30 centimeters and contain several seeds (Fig. **24.3**).

5. Seeds: The seeds are large, hard, and typically oblong or kidney-shaped. They can be white, beige, or brown in color, depending on the variety.

Distribution:

Africa, North America, Oceania, South America; Asia-China, India, Philippines, Malaysia, Taiwan, and Japan (USDA-ARS, 2016).

Propagation: Seed

Chemical constituents:

The seeds of *C. ensiformis* have higher levels of crude protein, crude lipid, and minerals such as sodium (Na), potassium (K), calcium (Ca), magnesium (Mg), phosphorus (P), iron (Fe), and manganese (Mn) compared to *C. gladiata*. The major proportion of seed proteins in both species consists of albumins and globulins (Rajaram *et al.*, 1992).

Fig. (24.3). Pods of *C. ensiformis* (PC: Pushpavati Bagul).

Recipe:

Ingredients: Pods, oil, mustard, garlic, onion, turmeric, hing, mix masala, goda masala, salt, coriander, *etc.*

Method: To prepare the dish, start by washing the pods and cutting them into two halves. Remove the veins/threads and discard the inner fleshy parts. Cut the pods into small pieces. Next, heat oil in a pan and add mustard, garlic, and onion. Sauté them until translucent. Add turmeric, hing, and mix masala, and then add the vegetables. Sprinkle goda masala and salt. Pour some water and cook for 10-15 minutes. Finally, garnish the dish with coriander and it is ready to serve.

Uses:

1. Protein Source: *C. ensiformis* seeds are rich in protein, making them a valuable source of plant-based protein (Brücher, 1989).

2. Dietary Fiber: The seeds contain dietary fiber, which can aid in digestion and promote a healthy digestive system (Akpapunam and Sefa-Dedeh 1997).

3. Antioxidant Properties: *C. ensiformis* seeds contain flavonoids and phenolic compounds, which have antioxidant properties. These compounds can help protect against oxidative stress and may have potential health benefits (Akpapunam and Sefa-Dedeh 1997).

4. Antimicrobial Effects: Certain compounds in *C. ensiformis* have demonstrated antimicrobial activity against certain bacterial and fungal strains (Akpapunam and Sefa-Dedeh 1997).

Canavalia gladiata (Jacq.) DC.

Botanical name: *Canavalia gladiata*

Family: Fabaceae

Local name: Sword Bean

Vernacular name:

- **Common name:** Jack bean, Jamaican horse bean, Scimitar bean, Sword bean
- **Assamese:** Kamtal urahi
- **Hindi:** Makkhan sem
- **Kannada:** Shimbe avare
- **Malayalam:** Vaal payara
- **Manipuri:** Tebi
- **Marathi:** Abai
- **Sanskrit:** Aasishimbi, Mahashimbi
- **Urdu:** Makkhan sem

Season: Flowering: July–Sept. Fruiting: October

Parts used: All parts

Characteristics:

1. Plant Habit: *C. gladiata* is a vine-like plant with slender stems that can climb or trail along the ground.

2. Leaves: The leaves are compound, trifoliate, and alternate. Each leaflet is lanceolate or ovate in shape and has a smooth or slightly hairy surface (vadivel *et al.*, 2010, Xia *et al.*, 2017).

3. Flowers: The flowers are typically large, showy, and papilionaceous (butterfly-shaped). They are usually white, pink, or purple in color and borne on long stalks.

4. Pods: The pods are elongated, cylindrical, and somewhat curved. They can grow up to several inches in length and contain several seeds (Vadivel *et al.*, 2004; Ekanayake, 2001) (Fig. **24.4**).

Fig. (**24.4**). Pods of *C. gladiata* (PC: Rakesh Sharma).

5. Seeds: The seeds are typically large, smooth, and oval-shaped. They can be white, cream, brown, or black, depending on the variety.

Distribution: Tropical Asia and Africa

Propagation: Seed

Chemical constituents:

Proteins-albumins and globulins, carbohydrates- starch and dietary fiber, flavonoids, lectins, Phytosterols- β-sitosterol, alkaloids, minerals- potassium, magnesium, calcium, phosphorus, and sulphur (Ekanayake *et al.*, 1999).

Recipe:

Ingredients: Pods, oil, mustard, garlic, onion, turmeric, hing, mix masala, goda masala, salt, coriander.

Method: First, wash the pods and cut them into halves. Remove the veins/threads and discard the inner fleshy parts. Then, chop the pods into small pieces. Heat oil in a pan and add mustard, garlic, and onion. Sauté them until translucent. Next, add turmeric, hing, and mixed masala, followed by the vegetables. Add goda masala and salt to taste. Sprinkle some water and cook the mixture for 10-15 minutes. Finally, garnish with fresh coriander before serving.

Uses:

1. Protein Source: *C. gladiata* seeds are rich in protein, making them a valuable source of plant-based protein (Pradeep *et al.*, 2014; Martinez, 2019).

2. Carbohydrates and Dietary Fiber: The seeds contain carbohydrates and dietary fiber, providing energy and promoting healthy digestion (Yamauchi *et al.*, 1987; Martinez, 2019).

3. Antioxidant Activity: Plant seeds contain flavonoids, which possess antioxidant properties. These compounds can help neutralize harmful free radicals and protect the body against oxidative stress (Yamauchi *et al.*, 1987; Martinez, 2019).

4. Anti-inflammatory Effects: Certain compounds found in *C. gladiata*, such as flavonoids, may exhibit anti-inflammatory properties. These effects can help reduce inflammation and potentially alleviate associated symptoms (Pradeep *et al.*, 2014; Martinez, 2019).

5. Hypoglycemic Activity: *C. gladiata* extracts have shown potential hypoglycemic effects in animal studies, suggesting a possible role in managing blood sugar levels and diabetes (Yamauchi *et al.*, 1987; Martinez, 2019).

6. Antimicrobial Properties: Seeds have been investigated for their antimicrobial activity against certain bacterial and fungal strains. These properties may contribute to its traditional use in folk medicine for treating microbial infections (Pradeep *et al.*, 2014; Martinez, 2019).

7. All parts of the plant have been used as a crude drug for the treatment of vomiting, abdominal dropsy, kidney-related lumbago, asthma, obesity, stomachache, dysentery, coughs, headache, intercostal neuralgia, epilepsy, schizophrenia, inflammatory diseases, and swellings (Yamauchi *et al.*, 1987; Martinez, 2019).

Cassia fistula L.

Botanical name: *Cassia fistula*

Family: Fabaceae

Local name: Bahava, Golden Shower

Vernacular name:

- **Common name:** Amaltas, Golden shower tree, Indian Laburnum
- **Hindi:** Amaltas
- **Kannada:** Kakke, Aaragu, Aaragina, Konde
- **Manipuri:** Chahui
- **Tamil:** Konrai
- **Malayalam:** Vishu konnai
- **Marathi:** Bahava
- **Mizo:** Ngaingaw
- **Bengali:** Sonali, Bandarlati, Amultas
- **Urdu:** Amaltas
- **Gujarati:** Garmalo
- **Sanskrit:** Aaragvadh, Raajavriksha Shampaak, Chaturangul
- **Oriya:** Sunari

Season: Leaf Fall: March- April, Flowering: April- June, Fruiting: December- April

Parts used: All parts

Characteristics:

1. Tree: *C. fistula* is a medium to large-sized deciduous tree that can reach a height of 10-20 meters (Rajagopal *et al.*, 2013).

2. Leaves: The leaves are pinnately compound, alternate, and arranged in an alternate pattern along the branches. Each leaf is composed of several pairs of leaflets, with a terminal leaflet at the end (Danish *et al.*, 2011; Neelam *et al.*, 2011; Rajagopal *et al.*, 2013).

3. Leaflets: The leaflets are elliptical or ovate in shape, with a smooth margin. They are usually about 10-20 centimeters long and 5-10 centimeters wide (Danish *et al.*, 2011; Neelam *et al.*, 2011; Rajagopal *et al.*, 2013).

4. Flowers: The tree produces large, showy, yellow flowers that hang in long

drooping clusters called racemes. Each flower has five petals, with the largest petal being the bottommost one (Danish *et al.*, 2011; Neelam *et al.*, 2011; Rajagopal *et al.*, 2013) (Fig. **24.5**).

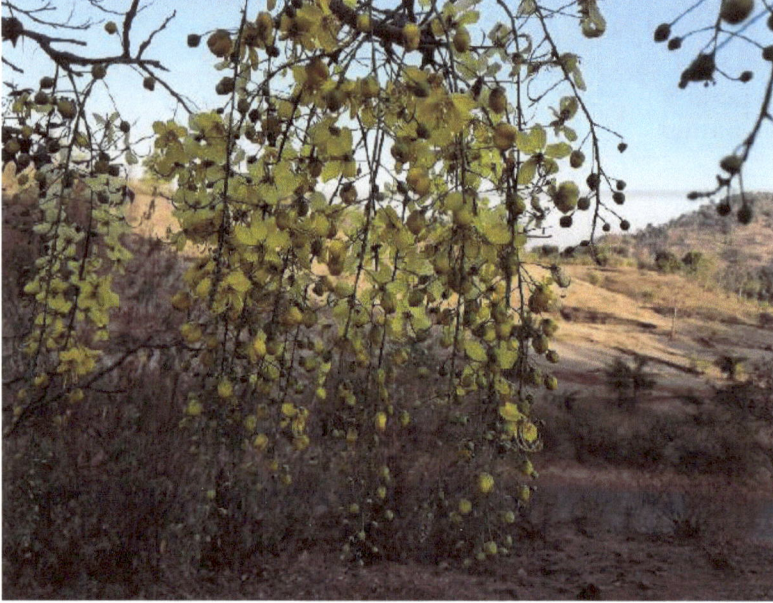

Fig. (**24.5**). **Inflorescence of *C. fistula*** (PC: Tukaram Kanoja).

5. Fruits: The fruits are long, cylindrical pods that hang in clusters. They are dark brown in color and can grow up to 30-60 centimeters in length. When mature, the pods become dry and brittle (Danish *et al.*, 2011; Neelam *et al.*, 2011; Rajagopal *et al.*, 2013).

6. Seeds: Inside the mature pods, there are numerous flat, black seeds embedded in a sticky pulp (Danish *et al.*, 2011; Neelam *et al.*, 2011; Rajagopal *et al.*, 2013).

Distribution:

India: Andhra Pradesh, Assam, Bihar, Kerala, Madhya Pradesh, Maharashtra, Odisha, Uttar Pradesh.

Propagation: Seeds, vegetative (Siddiqua *et al.*, 2008).

Chemical constituents:

Anthraquinones, flavonoids, flavon-3-ol derivatives, alkaloids, glycosides, tannin, saponin, terpenoids, reducing sugar, and steroids.

Recipe:

The young leaves and flower buds are cooked as a vegetable.

Uses:

1. Young Leaves: The young leaves of *C. fistula* are sometimes consumed as a leafy vegetable. They can be used in salads, stir-fries, or cooked as a side dish. The leaves are rich in nutrients and have a slightly bitter taste (Mondal, 2014; Mwangi *et al.*, 2021).

2. Flower Buds: In certain culinary traditions, the unopened flower buds are used as a vegetable. They can be cooked and consumed in various ways, such as stir-frying or adding to curries. The flower buds are often described as having a mildly sweet and slightly bitter flavor (Mondal, 2014; Mwangi *et al.*, 2021).

3. Medicinal Use: The fruit pulp and bark have been used in traditional medicine for their medicinal properties. They are known to have laxative, antipyretic, antifungal, and antibacterial effects. The fruit pulp is often used as a natural remedy for constipation (Mondal, 2014; Mwangi *et al.*, 2021).

4. Ornamental Plant: It is widely cultivated as an ornamental tree due to its beautiful golden-yellow flowers. It is commonly planted in gardens, parks, and along streets for its aesthetic appeal (Mwangi *et al.*, 2021).

5. Timber: The wood is durable and has been used for making furniture, cabinetry, and other wooden crafts. It is also used for construction purposes (Mondal, 2014; Mwangi *et al.*, 2021).

6. Dye: The flowers are used to extract a yellow dye. This dye has been traditionally used for coloring textiles and fabrics (Mondal, 2014; Mwangi *et al.*, 2021).

7. Bee Forage: The flowers attract bees, making them a valuable source of nectar for honey production. Beekeepers often plant these trees to support bee populations and honey production (Mondal, 2014; Mwangi *et al.*, 2021).

Cassia tora **L.**

Botanical name: *Cassia tora* L.

Family: Fabaceae (Sub family: Caesalpinioideae)

Local name: Takala

Vernacular name:

- **Hindi:** Chakavat, Charota, Chakvad
- **Bengali & Oriya:** Chakunda
- **Gujarati:** Kawaria
- **Canarese:** Gandutogache
- **Malayalam:** Chakramandrakam, takara
- **Sanskrit:** Chakramarda, Dadmari, Dadrughra, Taga
- **Tamil:** Tagarai
- **Telugu:** Chinnakasinda

Season: In India, it occurs as wasteland rainy season weed.

Parts used: Roots, leaves, and seeds.

Characteristics (Telrandhe and Gunde 2022):

1. Habit: *C. tora* is an annual herbaceous plant that grows erect or ascending, reaching a height of about 30-90 cm.

2. Leaves: The leaves are compound and alternate. Each leaf is a pinnately compound, consisting of 3 pairs of leaflets and a terminal leaflet. The leaflets are ovate or lanceolate in shape, with a smooth margin and a pointed tip. The leaflets are about 1.5-3 cm long (Fig. **24.6**).

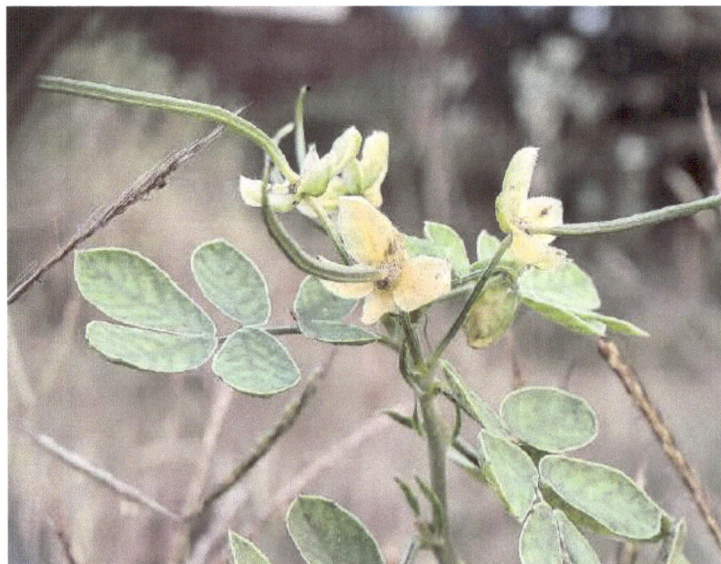

Fig. (24.6). Flowers of *C. tora* (PC: Tukaram Kanoja).

3. Flowers: The flowers are yellow in color and arranged in axillary racemes. Each flower has five petals, with the two lower petals larger and often marked with dark red or purple streaks.

4. Fruits: The fruits are cylindrical pods that are initially green but turn dark brown when mature. The pods are about 2-3 cm long and contain several small seeds.

5. Seeds: The seeds are small, shiny, and typically dark brown or black in color. They are oblong or kidney-shaped.

Distribution:

C. tora is widely distributed in various parts of the world. It is found abundantly in regions such as Afghanistan, India, Nigeria, China, Pakistan, Myanmar, Nepal, and Bhutan. In Nepal, it is cultivated in the Himalayan regions at an elevation of around 1400 meters. It has a broad distribution throughout India, Sri Lanka, West China, and tropical regions, with a particular presence in forested and tribal areas.

Propagation: By seeds

Chemical constituents:

Obtusin, chryso-obtusin, chryso-obtusin-2-O-β-D-glucoside, obtusifolin and chryso-obtusifolin-2-O- β -D-glucoside (Raghunathan *et al.*, 1974), 1, 3, 5 trihydroxy 6, 7 dimethoxy-2-methyle anthraquinone, leucopelargonidine and β-Sitosterol.

Recipe:

Ingredients: Younger leaves, finely chopped onion, 5-6 garlic cloves, salt, green chilies, curry leaves, Asafoetida, mustard seeds, turmeric, *etc.*

Method: Peel the stalks of the leaves and wash the curry leaves. Chop them finely. Heat oil in a pan and add asafoetida, mustard seeds, curry leaves, crushed garlic cloves, and finely chopped onion. When the onion turns slightly yellow, add the chopped vegetables, chili, turmeric powder, and salt to taste. Lightly beat the water, cover the pan, and cook over low heat.

Uses:

1. High in Protein: *C. tora* seeds are a good source of protein. Protein is essential for various functions in the body, including tissue repair, muscle development, and hormone production (Perry, 1980; Horvath, 1992).

2. Rich in Fiber: Seeds are high in dietary fiber, which aids in digestion and promotes bowel regularity. Fiber can also help regulate blood sugar levels and promote a feeling of fullness, making it beneficial for weight management (Zahra *et al.*, 2000; Harrison *et al.*, 2003).

3. Nutrient Content: The seeds contain essential nutrients such as vitamins, minerals, and antioxidants. These include calcium, iron, phosphorus, magnesium, and vitamin C (Perry, 1980; Horvath, 1992).

4. Food and Beverage Industry: Seeds are used as a food ingredient in various cuisines. They are commonly used in curries, soups, and stews. The seeds are also roasted and ground to make a coffee-like beverage (Zahra *et al.*, 2000; Harrison *et al.*, 2003).

5. Potential Antioxidant Activity: Seeds contain natural antioxidants that help protect the body against oxidative stress caused by free radicals. Antioxidants play a crucial role in maintaining cellular health and reducing the risk of chronic diseases (Zahra *et al.*, 2000; Harrison *et al.*, 2003).

6. Medicinal: It is widely used in traditional medicine systems for its medicinal properties. It shows laxative, diuretic, and purgative effects. It is used to treat constipation, skin diseases, eye infections, glaucoma, hypertension, skin disease, ringworm, leprosy, flatulence, colic, dyspepsia, constipation, cough, itch, and liver disorders (Perry, 1980; Horvath, 1992; Zahra *et al.*, 2000; Harrison *et al.*, 2003).

7. Agricultural: It is used in agriculture as a natural pesticide and weed suppressant. It has insecticidal properties and can help control pests and diseases in crops. The seeds are also used as a natural fertilizer (Zahra *et al.*, 2000; Harrison *et al.*, 2003).

8. Industrial: Seeds contain a high amount of polysaccharides, which have applications in various industries. They are used in the textile industry for printing and dyeing fabrics. The seeds are also used in the paper and adhesive industries (Zahra *et al.*, 2000; Harrison *et al.*, 2003).

9. Traditional: In traditional practices, *C. tora* possesses purifying and cleansing properties. It is used in rituals and ceremonies for spiritual purposes (Zahra *et al.*, 2000; Harrison *et al.*, 2003).

Leucaena leucocephala (Lam.) de Wit

Botanical name: *Leucaena leucocephala*

Family: Fabaceae

Local name: Safed babool

Vernacular name:

- **Common name:** Wild tamarind, Horse tamarind, White Babool, Leucaena, Lead tree
- **Hindi:** Safed babool
- **Bengali:** Subabul
- **Mizo:** Japan-zawngtah
- **Manipuri:** Chigong lei angouba

Season: July-October

Parts used: Young leaves and seed pods

Characteristics:

1. Growth Habit: *L. leucocephala* is a small to medium-sized tree or shrub that can reach a height of 6 to 15 meters (20 to 50 feet). It has an upright and spreading growth habit (Chiou *et al.*, 2013).

2. Leaves: The leaves of *L. leucocephala* are compound and fern-like in appearance. Each leaf is composed of many small leaflets arranged alternately along a central stalk (rachis). The leaflets are small and elongated, typically measuring about 1-3 centimeters (0.4-1.2 inches) in length (Chiou *et al.*, 2013).

3. Flowers: The tree produces small, fragrant flowers that are typically white or cream-colored. The flowers are arranged in spherical clusters called "heads" or "pom-poms" at the ends of branches. These flower clusters can be quite showy and attract pollinators (Hwang *et al.*, 2010; Pandey and Kumar, 2013) (Fig. **24.7**).

4. Fruits: After the flowers are pollinated, *L. leucocephala* develops seed pods. The pods are flat and elongated, measuring around 10-20 centimeters (4-8 inches) in length. They are usually light brown in color and contain several small, oblong seeds (Hwang *et al.*, 2010; Pandey and Kumar, 2013; Ngongolo *et al.*, 2014; Batisteli *et al.*, 2020).

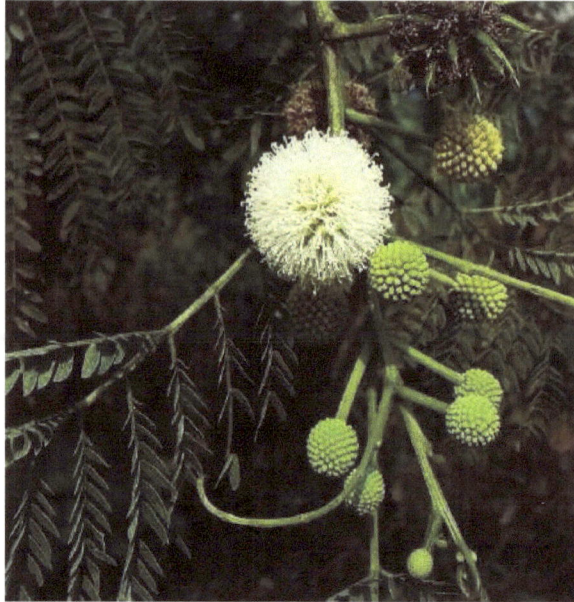

Fig. (24.7). Inflorescence of *L. leucocephala* (PC: Apurva Shankar Chonde).

5. Bark and Trunk: The bark of *L. leucocephala* is typically woody, cylindrical, branched, smooth and grayish-brown in color. As the tree matures, the bark becomes rougher and develops fissures (Chiou *et al.*, 2013).

6. Growth Pattern: *L. leucocephala* has a fast growth rate and can spread through both seeds and root suckers. It has a tendency to form dense thickets or stands, especially in disturbed or open areas.

Distribution:

Africa, Bhutan, Cambodia, Central America, Europe, India, Indonesia, Iran, Iraq, Japan, Laos, Malaysia, Pakistan, Philippines, Sri Lanka, Taiwan, Thailand, Vietnam, *etc.*

Propagation: Seeds, Stem cuttings

Chemical constituents:

1-octadecyne 5-octadecene, alkaloid, allyl hexadecyl ester, cardiac glycosides, Cis-coumaric acid, ficaprenol-11 (polyprenol), flavonoids, glycosides, hexatriacontane, lupeol, octacosane, oxalic acid, pheophorbide, a methyl ester pheophytin-a, saponins squalene tannins tetratetracontane, trans-coumaric acid (Wheeler *et al.*, 1996; Zayed *et al.*, 2018)

Recipe:

Young leaves, pods, and flower buds can be consumed raw or steamed. They can also be used in soups, paired with rice, or mixed with chilies and other spices for added flavor. The young pods are commonly cooked as a vegetable, and the roasted seeds can be used as a coffee substitute.

Uses:

1. Livestock Feed: The foliage of *L. leucocephala* is highly nutritious and serves as an excellent forage for livestock. It is particularly valuable as a feed source for ruminant animals due to its high protein content. The leaves and young shoots are palatable to cattle, goats, and sheep (González-García *et al.*, 2009).

2. Soil Improvement: *L. leucocephala* is a nitrogen-fixing plant, meaning it has the ability to convert atmospheric nitrogen into a usable form for plants. This trait enhances soil fertility and can improve the growth of other plants in the vicinity. It is often used as a green manure crop or as a cover crop to replenish nitrogen levels in agricultural fields (Kumar *et al.*, 1998; Isaac *et al.*, 2003).

3. Agroforestry: *L. leucocephala* is commonly incorporated into agroforestry systems due to its multiple benefits. Its fast growth and nitrogen-fixing abilities make it suitable for alley cropping or as a shade tree in agricultural fields. The tree provides shade to crops, enriches the soil, and its pruned foliage can be used as livestock feed (Yousif *et al.*, 2020).

4. Timber and Firewood: The wood of *L. leucocephala* is lightweight, durable, and suitable for various purposes. It is often used for construction, furniture-making, and fencing. Additionally, the tree produces a significant amount of firewood, providing a renewable energy source for cooking and heating (Yusuff *et al.*, 2019; Bageel *et al.*, 2020; Yousif *et al.*, 2020; De Angelis *et al.*, 2021; Ibrahim *et al.*, 2021).

Mimosa pudica L.

Botanical name: *Mimosa pudica*

Family: Fabaceae

Local name: Touch Me Not, Lajalu

Vernacular names:

- **Assamese:** Adori-bon, Lajuki-lata

- **Sanskrit:** Lajja
- **English:** Sensitive plant
- **Hindi:** Laajvanti and Chhui-mui
- **Bengali:** Lajjabati
- **Telugu:** Attapatti and Peddanidrakanni
- **Tamil:** Tottaaladi and Thottalchnungi
- **Kannada:** Lajja, Nachika and Mudugu-davare
- **Malayalam:** Tintarmani

Season: May to September

Parts used: All parts

Characteristics (Saraswat *et al.*, 2012):

1. Habit: The plant has a prostrate to semi-erect growth habit, often forming low, spreading clumps.

2. Leaves: Leaves are alternately arranged along the stems. Bipinnately compound leaves, meaning each leaf is composed of multiple pairs of leaflets. Each leaflet is small, usually oblong or obovate, with a pointed tip (Fig. **24.8**).

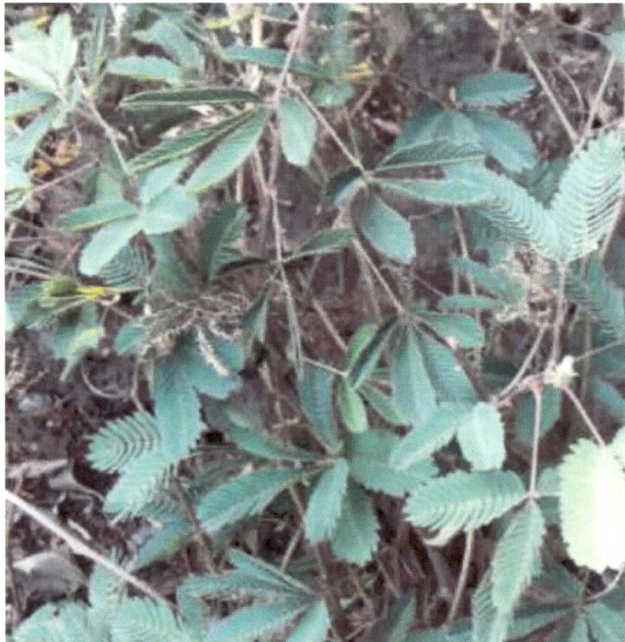

Fig. (24.8). Leaves of *M. pudica* (PC: Apurva Shankar Chonde).

3. Thigmonastic Movement: One of the most distinctive features of *M. pudica* is its thigmonastic movement, commonly known as "sensitive" or "sleep" movements. When touched or disturbed, the leaflets fold together and the entire leaf droops, giving the appearance of wilting. This is a defense mechanism to deter herbivores.

4. Stem: *M. pudica* has herbaceous, somewhat woody stems. The plant exhibits a branching growth habit.

5. Flowers: The flowers are small and arranged in spherical pink or purple globose heads. Bisexual, with both male and female reproductive organs in the same flower. The flowers have a small calyx and a tubular corolla.

6. Fruit: The fruit is a small, brown pod.

7. Roots: *M. pudica* typically has a fibrous and shallow root system.

Distribution:

It can also be found in Asian countries such as Singapore, Bangladesh, Thailand, India, Nepal, Indonesia, Taiwan, Malaysia, the Philippines, Vietnam, Cambodia, Laos, Japan and Sri Lanka. It has been introduced to many other regions and is regarded as an invasive species in Tanzania, South and Southeast Asia and many Pacific islands.

Propagation: Seeds, stem cutting

Chemical constituents:

Mimosine, alkaloids, amino acids, glycoside, fatty acids, flavonoids, phytosterol, and tannins. Crocetin, ascorbic acid, D-glucuronic acid, linoleic acid, D-xylose and B-sitosterols, tyrosin, vitexin, nor-epinephrine, d-pinitol, b-sitosterol, alkaloids-mimosine, terpenoids, flavonoids, glycosides, alkaloids, phenols, tannins, saponins, and coumarins, polyunsaturated fatty acid, sphingosine, adrenalin, Mimosine, 5-MeO-DMT, Norepinephrine, L-mimosine, Sphingosine, and isoorientin (Hassan *et al.*, 2019).

Recipe:

Ingredients: Younger leaves, finely chopped onion, 5-6 garlic cloves, salt, green chilies, curry leaves, Asafoetida, mustard seeds, turmeric, gram dal, *etc.*

Method: Peel the stalks of the leaves, wash one Judy leaves, and chop finely. Heat oil in the pan and fry asafoetida, mustard seed, curry leaves, crushed garlic cloves, and finely chopped onion. When the onion turns slightly yellow, add the chopped vegetables, chili, turmeric powder, soaked dal, and salt to taste, lightly beat water and cover and cook over low flame.

Uses:

1. Traditional Medicine: In traditional medicine systems, various parts of *M. pudica* are used to treat conditions such as diarrhea, dysentery, respiratory issues, and skin ailments (Khare, 2004; Chatterjee *et al.*, 2006).

2. Anti-inflammatory Properties: The plant shows anti-inflammatory properties, and extracts are sometimes used for conditions involving inflammation (Prajapati *et al.*, 2003; Behera *et al.*, 2006).

3. Ornamental Purposes: Indoor Plant: Due to its unique leaf movements and attractive foliage, *M. pudica* is sometimes grown as an indoor plant or as part of ornamental gardens (Behera *et al.*, 2006).

Mucuna pruriens **(L.) DC.**

Botanical name: *Mucuna pruriens*

Family: Fabaceae

Local name: Kuyari, Khaajkuiri

Vernacular name:

- **Sanskrit:** Atmagupta, Kapikacchu
- **Hindi:** Kiwanch
- **Marathi:** Khaajkuiri
- **Bengali:** Alkushi
- **Tamil:** Poonaikkaali
- **English:** Velvet bean, Cowitch, Cowhage
- **German:** Juckbohne, Itch bean
- **Portuguese:** Mad Bean
- **Malayalam:** Naykaranam
- **Thai language:** MahMui

Season: September to January

Parts used: young leaves, pods and seeds

Characteristics (US Forest Service, 2011):

1. Plant Type: *M. pruriens* is a climbing or trailing plant. It has long, twining stems that can grow several meters in length. The plant uses tendrils to climb and attach to structures or other plants.

2. Leaves: The leaves of *M. pruriens* are alternate and trifoliate, meaning they consist of three leaflets. The leaflets are ovate or elliptic in shape, and they have pointed tips and smooth margins. The leaves are green and have a velvety or hairy texture.

3. Flowers: The plant produces clusters of flowers that are borne on long stalks or racemes. The flowers are usually purple or lavender in color, but they can also be white or pink. Each flower has five petals and a tubular shape.

4. Seed Pods: *M. pruriens* produces elongated, cylindrical seed pods. The pods are covered in dense, bristly hairs, giving them a prickly or velvet-like texture. The color of the pods varies from green to brown as they mature (Fig. **24.9**).

Fig. (**24.9**). Pods of *M. pruriens* (PC: Apurva Shankar Chonde).

5. Seeds: Inside the seed pods, *M. pruriens* bears oval or kidney-shaped seeds. The seeds are usually brown or black in color and have a smooth surface. They are relatively large compared to other legume seeds.

6. Hairs and Trichomes: One unique characteristic of *M. pruriens* is the presence of stinging hairs or trichomes on the seed pods and leaves. These hairs contain a

chemical called serotonin, which can cause itching and irritation upon contact with the skin.

Distribution:

Africa, America, Asia, India, Pacific Islands.

Propagation: Seed

Chemical constituents:

L-DOPA (Levodopa), alkaloids, flavonoids, sterols, proteins, triterpenes, saponins, *etc.*

Recipe:

Ingredients: Tomatoes, mustard seeds, cumin seeds, curry leaves, chopped onion, crushed garlic-ginger, *etc.*

Method: For the vegetables, chop the onions, crush the garlic and ginger, and dice the tomatoes. In a pan, heat oil and add mustard seeds, cumin seeds, curry leaves, chopped onions, and crushed garlic and ginger. Cook them until the onions turn translucent and the tomatoes are cooked. Then, add the chopped curry leaves and water, followed by the beans. Cook the vegetables until they are tender. Finally, add salt to taste.

Uses:

1. Traditional Medicine: *M. pruriens* has a long history of use in traditional medicine systems, particularly in Ayurveda. It has been used to support overall well-being, promote vitality, and improve sexual health. Traditional uses also include managing nervous system disorders, promoting muscle strength, and supporting reproductive health (Jeyaweera, 1981; Sathiyanarayanan *et al.*, 2007).

2. Dopamine Support: *M. pruriens* is known for its high content of L-DOPA (Levodopa), a precursor to the neurotransmitter dopamine. L-DOPA is used as a medication for Parkinson's disease to help replenish dopamine levels in the brain. *M. pruriens* supplements or extracts with standardized L-DOPA content may support dopamine production and potentially offer neuroprotective effects (Misra and Wagner, 2007).

3. Male Sexual Health: *M. pruriens* has been used traditionally as an aphrodisiac and to support male sexual health. It may help promote healthy libido, enhance

fertility, and support healthy testosterone levels. The *M. pruriens* may have positive effects on sperm quality and motility (Rajeshwar *et al.*, 2005).

4. Antioxidant and Anti-inflammatory Effects: *M. pruriens* contains various antioxidants and anti-inflammatory compounds, such as flavonoids and other phytochemicals. These compounds help neutralize harmful free radicals and reduce inflammation in the body, potentially offering protective effects against oxidative stress and certain chronic conditions (Hishika *et al.*, 1981; Rajeshwar *et al.*, 2005).

5. Nutritional Value: *M. pruriens* seeds are a good source of protein, fiber, and various minerals. They can be consumed as a nutritious food source and may contribute to a balanced diet (Lampariello *et al.*, 2012).

Senna occidentalis L.

Botanical name: *Senna occidentalis*

Family: Fabaceae

Local name: Ran-takda, kasivda

Vernacular name:

- **Hindi:** Kasunda, Bari kasondi
- **Tamil:** Nattam takarai, Payaverai
- **Malayalam:** Mattantakara
- **Telugu:** Thangedu
- **Kannada:** Kolthogache, Aane chogate
- **Bengali:** Kalkashunda
- **Oriya:** Kasundri
- **Urdu:** Kasonji
- **Assamese:** Hant-thenga
- **Gujarati:** Kasundri
- **Sanskrit:** Kasamarda, Vimarda, Arimarda
- **Mizo:** Reng-an

Season: Rainy season

Parts used: Leaves, seeds

Characteristics (Calalb *et al.*, 2021):

1. Plant Habit: *S. occidentalis* is an annual or short-lived perennial herbaceous plant. It typically grows erect and can reach heights between 0.5 to 2 meters (1.6 to 6.6 feet).

2. Leaves: The leaves of *S. occidentalis* are pinnately compound, and alternately arranged along the stems. Each leaf is composed of several pairs of leaflets, usually ranging from 4 to 8 pairs. The leaflets are elongated and lanceolate in shape, with a pointed apex and smooth margins. They are typically dark green in color.

3. Flowers: The plant produces small, bright yellow flowers that are arranged in elongated clusters or racemes. Each flower has five petals, with the uppermost petal being larger and often slightly curved. The flowers have both male and female reproductive parts and exhibit typical characteristics of the Fabaceae family (Fig. **24.10**).

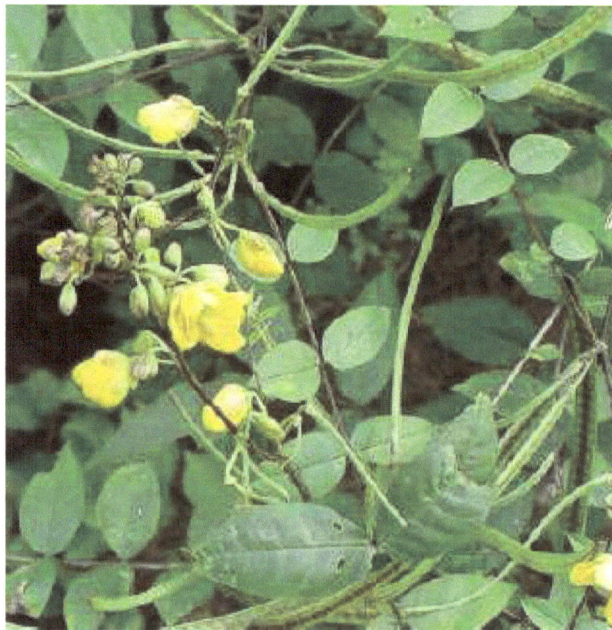

Fig. (24.10). Inflorescence of *S. occidentalis* (PC: Apurva Shankar Chonde).

4. Fruits: After pollination, *S. occidentalis* develops elongated, slender pods or legumes. The pods are initially green and turn dark brown or black when mature. They are cylindrical in shape and contain several seeds.

5. Seeds: The seeds of *S. occidentalis* are small, smooth, and shiny. They are typically black or dark brown in color and have a flattened or rounded shape.

Distribution:

North America- Florida, Texas, and Louisiana, Central America and the Caribbean- Mexico, Belize, Guatemala, Honduras, Nicaragua, Costa Rica, Panama, Cuba, Jamaica, and Puerto Rico. South America- Colombia, Venezuela, Guyana, Suriname, French Guiana, Brazil, Ecuador, Peru, Bolivia, Paraguay, and Argentina. Africa and Asia- Nigeria, Ghana, Sudan, Ethiopia, India, Sri Lanka, and Malaysia (Acevedo-Rodríguez and Strong, 2012).

Propagation: Seeds

Chemical constituents:

Anthraquinones- sennosides A, B, C, and D, Flavonoids- kaempferol, quercetin, and rutin, Alkaloids- catharticin, chrysophanic acid, and rhein, Triterpenes-betulinic acid and oleanolic acid, Sterols- β-sitosterol and stigmasterol. Volatile compounds- hydrocarbons, alcohols, aldehydes, and ketones

Recipe:

Ingredients: Younger leaves, finely chopped onion, 5-6 garlic cloves, salt, green chilies, curry leaves, Asafoetida, mustard seeds, turmeric, *etc.*

Method: Peel the stalks of the leaves, wash one bunch of leaves, and chop them finely. Heat oil in a pan and fry asafoetida, mustard seeds, curry leaves, crushed garlic cloves, and finely chopped onion. When the onion turns slightly yellow, add the chopped vegetables, chili, turmeric powder, and salt to taste. Lightly beat water, cover, and cook over low flame.

Uses:

1. Medicinal: *S. occidentalis* has a long history of use in traditional medicine systems. It is primarily known for its laxative properties. The seeds and leaves of *S. occidentalis* are often used to prepare herbal remedies for constipation relief. The active compounds called sennosides are present in the plant, which can stimulate bowel movements (Usha *et al.*, 2007).

2. Veterinary Purposes: *S. occidentalis* has also been used in veterinary medicine to treat digestive issues in animals. It is sometimes administered to livestock to alleviate constipation or as a vermifuge to expel intestinal worms (Usha *et al.*, 2007).

3. Agroforestry and Livestock Feed: In some regions, *S. occidentalis* is grown as part of agroforestry systems. The plant can provide shade for crops, act as a

windbreak, and improve soil fertility. Additionally, the leaves of *S. occidentalis* can be used as fodder for livestock, providing a source of nutrition (Usha *et al.*, 2007).

4. Green Manure: *S. occidentalis* is occasionally used as a green manure crop. When incorporated into the soil, its biomass can enhance soil fertility by adding organic matter and fixing nitrogen (Usha *et al.*, 2007).

5. Insecticide: Some studies have investigated the insecticidal properties of *S. occidentalis*. Certain compounds found in the plant have shown potential for use as natural insecticides or insect repellents (Usha *et al.*, 2007).

Sesbania grandiflora (L.) Pers.

Botanical name: *Sesbania grandiflora*

Family: Fabaceae

Local name: Hadaga

Vernacular name:

- **Hindi:** Gaach-munga, Hathya, Agasti
- **Manipuri:** Hawaiman
- **Marathi:** Shevari, Hatga
- **Tamil:** Sevvagatti, Muni
- **Malayalam:** Akatti
- **Telugu:** Avisha, Ettagise, Sukanasamu
- **Kannada:** Agasi, Agase, Chinnadaare, Arisina jeenangi
- **Bengali:** Buko, Bak
- **Urdu:** Agst
- **Gujarati:** Agathio
- **Sanskrit:** Varnari, Munipriya, Agasti, Drigapalaka

Season: September to January

Parts used: bark, leaves, flowers and roots

Characteristics (Bhat and Menon, 1971):

1. Habitat: *S. grandiflora* is native to tropical and subtropical regions of Asia, including India, Sri Lanka, Myanmar, Thailand, Malaysia, and Indonesia. It is often cultivated in home gardens, orchards, and agroforestry systems.

2. Plant Habit: *S. grandiflora* is a deciduous tree that can reach heights of 3 to 8 meters (10 to 26 feet). It has a slender trunk with a smooth grayish-brown bark.

3. Leaves: The leaves of *S. grandiflora* are pinnately compound, meaning they are composed of multiple leaflets arranged in pairs along the stem. Each leaf typically has 20 to 40 pairs of small leaflets that are lanceolate or oblong in shape. The leaflets are light green in color and have smooth margins.

4. Flowers: The tree produces showy, large flowers that are generally white or pink in color. The flowers are borne in long, hanging racemes, with each raceme containing several individual flowers. The petals are clawed, and the uppermost petal is larger and often marked with colorful patterns. The flowers are tubular and have an appearance that resembles hummingbirds, hence the common name "hummingbird tree" (Fig. **24.11**).

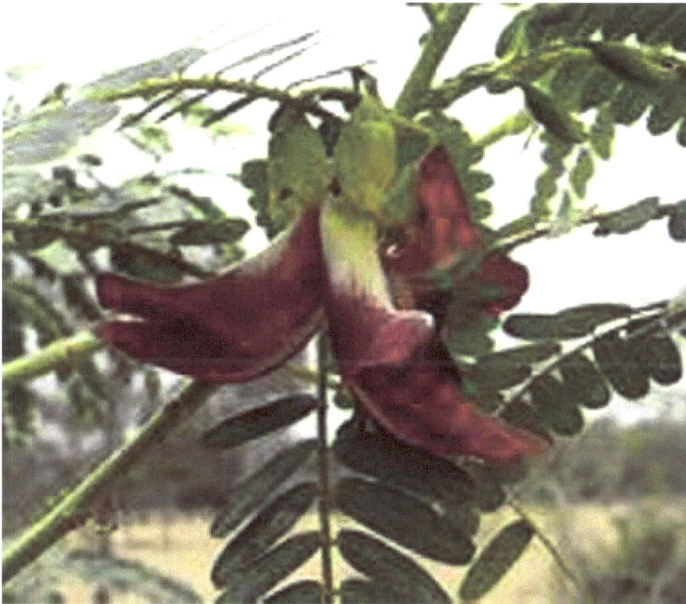

Fig. (**24.11**). Flowers of *S. grandiflora* (PC: Anil Phutane).

5. Fruits: After successful pollination, *S. grandiflora* develops long, slender pods or legumes. The pods are flat and elongated, reaching lengths of about 30 to 60 centimeters (12 to 24 inches). They are green when young and turn brown as they mature. Inside the pods are numerous small, oblong seeds.

Distribution: It is native to Southeast Asia (Malaysia, Philippines, and Brunei) to Northern Australia, and is cultivated in many parts of India and Sri Lanka.

Propagation: Seed, Cuttings of half-ripe wood.

Chemical constituents:

Alkaloids, flavonoids, glycosides, tannin, anthraquinone, steroids, pholobatannins, and terpenoids. Isovestitol, medicarpin, sativan (iso flavonoids) and betulinic acid (tannin substance), phenolics, tannins, triterpenoids, sterols, polyphenols, carotenoids, vitamins (vitamin A, C, riboflavin, nicotinic acid), and minerals like zinc and selenium (Zarena *et al.*, 2014; Velusamy *et al.*, 2016; Mohiuddin *et al.*, 2019).

Recipe:

Ingredients: Hadagya flowers, onion, garlic, chili, asafoetida, cumin, oil, *etc.*

Method: Vegetables should be washed and chopped. In a pan, fry the chopped onion in oil. Then add garlic, cumin or cumin powder, and wet chilies. Next, add the chopped vegetables and cook.

Uses:

1. Edible Parts: The young leaves, flowers, and tender pods of *S. grandiflora* are commonly used as a vegetable in many cuisines. They can be cooked and incorporated into various dishes, including soups, stir-fries, curries, and salads. The flowers are often used in traditional recipes for their mild flavor and nutritional value (Thissera *et al.*, 2020).

2. Nutritional Value: The leaves and pods of *S. grandiflora* are rich in vitamins, minerals, and dietary fiber. They are a good source of protein, vitamin C, calcium, iron, and other essential nutrients, making them a valuable addition to a balanced diet (Hussain *et al.*, 2014).

3. Medicinal: *S. grandiflora* has a long history of use in traditional medicine systems. Different parts of the tree, such as the leaves, flowers, bark, and roots, which possess medicinal properties. They are used in remedies for various ailments, including digestive disorders, respiratory issues, skin problems, and inflammation (Gomase *et al.*, 2012; Rajagopal *et al.*, 2016; Aye *et al.*, 2019).

4. Fodder and Livestock Feed: *S. grandiflora* is valued as a fodder plant for livestock due to its high nutritional content. The leaves and pods can be fed to cattle, goats, and other grazing animals, providing them with a nutrient-rich food source (Rajagopal *et al.*, 2016).

5. Soil Improvement: *S. grandiflora* is known as a nitrogen-fixing plant, meaning it has the ability to convert atmospheric nitrogen into a usable form that enriches the soil. It is often used in agroforestry systems, intercropping, and reforestation efforts to improve soil fertility and structure (Rajagopal *et al.*, 2016).

6. Environmental Benefits: *S. grandiflora* has the potential to provide shade, windbreak, and erosion control due to its fast growth and dense foliage. It is sometimes planted for landscaping and environmental restoration purposes (Rajagopal *et al.*, 2016).

Simithia bigemina Dalzell.

Botanical name: *Simithia bigemina*

Family: Fabaceae

Local name: Lahan kawala

Vernacular name:

- **Common name:** Double Paired Smithia
- **Marathi:** Lahan kawala, Nalagi

Season: Flowering and fruiting: September-February

Parts used: Leaves

Characteristics (Balan and Pradeep, 2017):

1. Habit: The habit of *S. bigemina* is that of slender prostrate herbs, indicating their growth close to the ground with stems creeping along the surface.

2. Stems: The stems of *S. bigemina* are densely covered in brown hairs, imparting a rough texture to them.

3. Leaves: The leaves exhibit an alternate arrangement, meaning they attach to the stem one at a time. Each leaf consists of only 2 leaflets, each measuring approximately 0.5-0.8 cm in length and 0.2-0.3 cm in width. These oblong-shaped leaflets are longer than they are wide, with roughly parallel sides. The undersides of the leaflets are characterized by a covering of soft hairs (villous).

4. Flowers: The flowers of *S. bigemina* are arranged in racemes, elongated clusters growing from the leaf axils and the tip of the stem. Each raceme comprises 4-5 individual flowers. The pedicels, supporting each flower, are thin

and thread-like (filiform). Bracts, small leaves at the base of the flower stalks, take on a lanceolate shape, resembling a spearhead with a wider base and pointed tip (Fig. **24.12**).

Fig. (24.12). Flowers of *S. bigemina* (PC: Apurva Shankar Chonde).

5. Calyx: The calyx, forming a cup-like structure around the base of the flower, measures 0.3-0.4 cm long. It is divided into two lips: the upper lip with 2 lobes and the lower lip with 3 lobes. The calyx is covered in stiff yellow bristles.

6. Petals: Petals of *S. bigemina* are yellow with bluish striations, contributing to a striped pattern. They range in size from 0.5-0.6 cm long. The wing petals, two lateral petals, exhibit transverse folds, characterized by wrinkles running across their width.

7. Stamens: Stamens are arranged in two sets of 5 each, totaling 10 stamens. They are enclosed in a tube-like structure called the staminal sheath, measuring 0.4-0.45 cm long.

8. Ovary: The ovary of *S. bigemina* is curved or twisted in shape.

9. Pods: It contains seeds that are 0.3-0.4 cm long and consist of 3-4 segments. Each segment is roughly circular (orbicular) and approximately 0.1 cm in diameter. The segments are twisted, contributing to the distinctive appearance of the fruit.

Distribution:

Pakistan and India (Andhra Pradesh, Goa, Gujarat, Karnataka, Kerala, Madhya Pradesh, Maharashtra, Odisha, Rajasthan, Tamil Nadu and Uttar Pradesh).

Propagation: Seed

Recipe:

Ingredients: Younger leaves, finely chopped onion, 5-6 garlic cloves, salt, green chilies, curry leaves, Asafoetida, grated coconut, mustard seeds, turmeric, and gram dal.

Method: Wash leaves and chop finely. Heat oil in the pan and fry asafoetida, mustard seed, curry leaves, crushed garlic cloves, and finely chopped onion. When the onion turns slightly yellow, add the chopped vegetables, chili, turmeric powder, soaked dal, salt to taste, lightly beat water and cover, and cook over low heat. Once the vegetables cook, add wet coconut on top.

Uses:

1. Culinary: Some tribal peoples use tender leaves as vegetables (Gawade and Kode 2020).

2. Fodder: The tender leaves and stems of *S. bigemina* can be used as forage for livestock, particularly goats and sheep, in areas where grazing options are limited (Gawade and Kode 2020).

3. Ornamental: While not widely cultivated, the delicate yellow flowers and unique twisted pods of *S. bigemina* could hold some appeal for wildflower enthusiasts or those seeking plants for natural landscaping (Gawade and Kode 2020).

Smithia sensitiva **Aiton**

Botanical name: *Smithia sensitiva*

Family: Fabaceae

Local name: Kavala

Vernacular name

• **Bengali:** Nalakashina

- **Hindi:** Odabirni
- **Marathi:** Lajalu Kavla
- **Other:** Sensitive Smithia

Season: Flowering: August-October.

Parts used: Leaves and young pods

Characteristics (Balan and Pradeep 2017):

1. Habit: *S. sensitiva* is a small annual herb with a prostrate or semi-erect growth habit. It usually grows up to about 15-30 centimeters in height.

2. Leaves: The leaves of *S. sensitiva* are compound and consist of three leaflets. The leaflets are oblong to elliptical in shape and have smooth margins. They are typically green in color and arranged alternately along the stem.

3. Inflorescence: The flowers of *S. sensitiva* are arranged in loose clusters or racemes at the ends of the branches. Each raceme typically bears several individual flowers.

4. Flowers: The flowers of *S. sensitiva* are small and showy, borne on long stalks. They are usually bright yellow in color, although variations with pale yellow or orange hues are also possible. The flowers are bilateral and have a distinctive shape, resembling a butterfly or pea flower. They have five petals, with the uppermost petal modified into a large, curved banner (Fig. **24.13**).

5. Fruits: After successful pollination, *S. sensitiva* produces small, elongated seed pods. These pods contain several seeds and turn brown or black when mature.

Distribution:

1. India: *S. sensitiva* is found throughout India. It occurs in various states across different regions, including North India, South India, Central India, and the Western Ghats. It can be found in states such as Maharashtra, Karnataka, Tamil Nadu, Andhra Pradesh, Telangana, Odisha, West Bengal, Bihar, Uttar Pradesh, Madhya Pradesh, and Rajasthan.

2. Other South Asian countries: *S. sensitiva* is also reported to occur in neighboring countries of South Asia, including Nepal, Bangladesh, and Sri Lanka.

Fig. (24.13). Flowering Twig of *S. sensitiva* (PC: Apurva Shankar Chonde).

Propagation: Seed

Chemical constituents: Flavonoids- kaempferol, quercetin, and their glycosides, alkaloids, triterpenes, and phenolic compounds.

Recipe:

Ingredients: Leaves, finely chopped onion, 5-6 garlic cloves, salt, green chilies, curry leaves, Asafoetida, mustard seeds, turmeric, and chilies.

Method: Wash the judy leaves thoroughly. Heat oil in a pan and fry asafoetida, mustard seeds, curry leaves, crushed garlic cloves, finely chopped onions, and chilies. Once the onions turn slightly yellow, add the chopped vegetables, chili, turmeric powder, and salt to taste. Lightly beat some water, cover the pan, and cook over low heat.

Uses:

1. Traditional Medicine: In traditional practices, *S. sensitiva* has been used for its potential medicinal properties. It possesses diuretic, anti-inflammatory, and antimicrobial properties. It is sometimes used to alleviate conditions such as urinary tract infections, rheumatism, and skin disorders (Udupa *et al.*, 2007; Sreena *et al.*, 2011).

2. Culinary: In certain regions, the tender shoots and young leaves of *S. sensitiva* are consumed as a vegetable. They are often cooked and used in traditional dishes such as stir-fries, curries, and soups (Chalil and Shinde 2021).

Tamarindus indica L.

Botanical name: *Tamarindus indica*

Family: Leguminosae

Local name: Chinch

Vernacular name:

- **Assamese:** Teteli
- **English:** Tamarind tree
- **Hindi:** Imli
- **Kannada:** Hunse
- **Malayalam:** Amlam
- **Tamil:** Puli

Season: All season plant, fruiting March to May

Parts used: Bark, leaves, flowers, fruits, seeds

Characteristics:

1. Size and Shape: Tamarind trees are large, evergreen or semi-deciduous trees that can reach heights of up to 25 meters (82 feet). They have a wide and spreading crown with a symmetrical, umbrella-like shape (Bhadoriya *et al.*, 2011).

2. Bark: The bark of the tamarind tree is dark brown or grayish-black and becomes rough and fissured as the tree matures. The surface of the bark can have a slightly scaly texture (Bhadoriya *et al.*, 2011).

3. Leaves: The leaves of *T. indica* are pinnately compound, meaning they consist of multiple leaflets arranged along a central axis or rachis. Each leaf typically has 10 to 20 pairs of leaflets. The leaflets are oblong or elliptical in shape, about 2 to 7 centimeters long, and have a glossy dark green color (Bhadoriya *et al.*, 2011).

4. Flowers: Tamarind trees produce small, fragrant flowers that are yellowish with reddish or orange veins. The flowers are borne in dense, drooping clusters or racemes, typically at the ends of branches (Bhadoriya *et al.*, 2011).

5. Fruit: The most recognizable feature of *T. indica* is its fruit, known as tamarind. The fruit is a large, brown pod that is 5 to 20 centimeters in length. The pod contains a sticky, brown pulp that surrounds the seeds. The pulp is sour and tangy, and it is used in various culinary preparations (Bhadoriya *et al.*, 2011) (Fig. **24.14**).

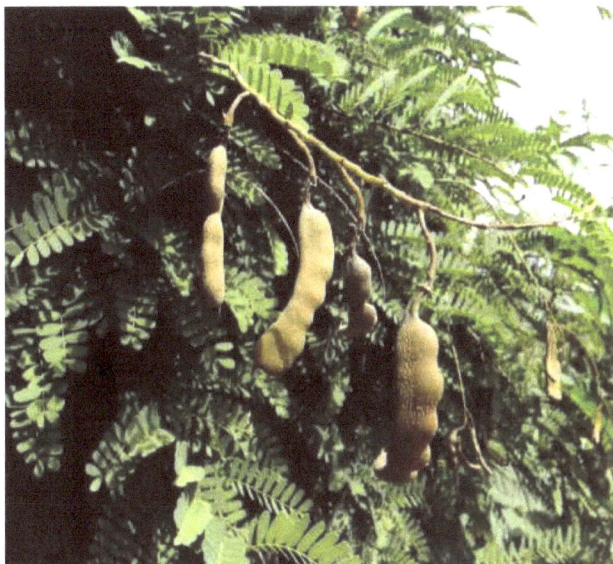

Fig. (**24.14**). Fruits of *T. indica* (PC: Apurva Shankar Chonde).

6. Seeds: Inside the tamarind fruit, there are several hard, shiny brown seeds. The seeds are somewhat flattened and have a smooth surface (Bhadoriya *et al.*, 2011).

Distribution:

Cameroon, China, India, Kenya, Malawi, Nigeria, Somalia, Southeast Asia, Tanzania, Taiwan and Zambia.

Propagation: By seed

Chemical constituents:

Tannins, polyphenols, organic acids, vitamin c, carbohydrates, volatile compounds, minerals- potassium, calcium, magnesium, phosphorus, and iron.

Recipe:

Ingredients: Young leaves, finely chopped onion, 5-6 garlic cloves, salt, green chilies, curry leaves, Asafoetida, mustard seeds, turmeric, and chilies.

Method: Wash a few young leaves of tamarind. Heat oil in a pan and fry asafoetida, mustard seeds, curry leaves, crushed garlic cloves, and finely chopped onions and chilies. When the onions turn slightly yellow, add the chopped vegetables, along with chili powder, turmeric powder, and salt to taste. Lightly beat some water, cover the pan, and cook over low heat.

In addition, the ripened fruits can be used in various culinary dishes.

Uses:

1. Culinary: The pulp of the tamarind fruit is used to add a tangy and sour flavor to dishes, sauces, chutneys, curries, marinades, and beverages. It is also a key ingredient in popular dishes like tamarind rice and tamarind-based soups (Benthall, 1933; Dalziel, 1937; Eggeling *et al.*, 1951; Little *et al.*, 1964).

2. Medicinal: Tamarind has been used in traditional medicine for centuries. It shows various health benefits. The pulp of the tamarind fruit is known for its laxative properties and is used to relieve constipation. Tamarind is also used to treat digestive disorders, fever, sore throat, and inflammation. It contains antioxidants and has antimicrobial properties (Sudjaroen *et al.*, 2005; Al-Fatimi *et al.*, 2007).

3. Industrial: Tamarind has industrial applications as well. Tamarind seed powder is used in textile and dye industries as a sizing material for cotton and silk fabrics. The seeds are also a source of tamarind kernel powder, which is used in the manufacturing of adhesives, glues, and thickening agents (Kulkarni *et al.*, 1993; Dagar *et al.*, 1995; George *et al.*, 1997).

4. Oral Health: Tamarind is used in oral care products. Its pulp and leaves have antimicrobial properties that help combat oral bacteria and improve oral hygiene. Tamarind is used in mouthwashes, toothpaste, and natural remedies for oral health issues like gum infections and bad breath (Dagar *et al.*, 1995).

5. Traditional Preservative: Tamarind extract and pulp have been traditionally used as natural preservatives for food. The high acidity and antimicrobial properties of tamarind help inhibit the growth of bacteria and fungi, thereby extending the shelf life of certain perishable foods (Dagar *et al.*, 1995).

6. Beverages: Tamarind is used to make refreshing beverages. Tamarind juice, often sweetened and diluted, is popular in many tropical regions. It is also used to make traditional drinks like "agua de tamarindo" in Latin America and "sherbet" in some parts of India (Anonymous 1982).

Vigna khandalensis (Santapau) Sundararagh. & Wadhwa

Botanical name: *Vigna khandalensis*

Family: Fabaceae

Local name: Ranmung

Season: Flowering and fruiting: August to October

Parts used: Seeds

Characteristics (Umdale *et al.*, 2018):

1. Growth habit: *V. khandalensis* is a tall and robust annual herb with an erect growth habit. The stems are straight, with 1-2 primary branches at the base. They are dark green with purple coloration and have a distinctive 5-angled shape. The stems are sparsely covered with yellowish white hairs measuring 0.6-1.2 mm in length. The leaf petioles are also sparsely covered with yellowish white hairs.

2. Leaves: The leaves of *V. khandalensis* are large and trifoliolate. They are dark green in color and sparsely strigose on both surfaces, with appressed bristly yellowish hairs. The leaflets have variable shapes, with a distantly wavy margin. The terminal leaflet is conspicuously trilobed and broadly ovate, while the lateral leaflets are 2-3 lobed and acute at the apex.

3. Stipules: The stipules of *V. khandalensis* are very large and foliaceous. They vary in size and have a sub-medifixed position. The stipules are ovate-cuspidate and sparsely ciliate. They are cordate at the base and obtuse at the apex, with cilia that are whitish in color.

4. Inflorescence: The inflorescence of *V. khandalensis* is axillary and consists of 15-25 flowers. The peduncles are long and sparsely covered with brownish-black retrorse bristly hairs. The flowers are arranged in axillary spicate racemes. The pedicels are short, and the bracts are broadly ovate and acute, but deciduous. The bracteoles are very large, oblong or obovate-oblong, and obtuse. They are veined and conceal the flower in bud.

5. Pods: The pods of *V. khandalensis* are linear and cylindrical in shape. They are brownish black or black when mature and separated between the seeds. The pods are densely pubescent with white or brown bristles. The aril is not developed, and the seed testa is rough (Umdale *et al.*, 2018) (Fig. **24.15**).

Fig. (24.15). Seeds of *V. khandalensis* (PC: Gaytri M. Chonde).

Distribution:

Western Ghats and the Deccan Plateau in India.

Flowering and fruiting- August to October

Propagation: Seed

Chemical constituents:

The seeds of *V. khandalensis* have approximately 48% carbohydrates, 24% proteins, and 5.5% lipids (Umdale *et al.*, 2018).

Recipe:

Ingredients: Sprouted mung beans, onion, garlic, ginger, tomato, turmeric powder, chili powder, coriander powder, cumin powder, garam masala, water, salt, oil.

Method: Heat the oil in a pressure cooker over medium heat and add the ginger, garlic, and onion. Sauté for a minute until the onion becomes translucent. Add the tomato and stir. Allow the tomato to cook down until it becomes soft, which should take about five minutes. Cover the cooker and stir occasionally during this process. Next, add the dry spice powders, including turmeric, chili, cumin, garam masala, and coriander powder. Fry them for a few seconds and then add the water. Mix everything thoroughly. Add the sprouted Mung beans and stir well. Pour in ½ a cup of water and add salt. Pressure cook this mixture for one whistle. Serve it hot with roti or rice.

Uses:

Both unripe and mature seeds of this species are cooked as a vegetable or eaten raw (Umdale *et al.*, 2018).

Vigna unguiculata (L.) Walp.

Botanical name: *Vigna unguiculata*

Family: Fabaceae

Local name: alasunda, chavali

Vernacular name:

- **Common name:** Blackeyed Bean, Cowpea, Blackeyed Pea
- **Bengali:** Ghagra
- **Hindi:** Lobiya
- **Kannada:** Alasabde, Alasundi
- **Marathi:** Alasunda, Chavali
- **Sanskrit:** Mahamasah, Rajamasah
- **Tamil:** Kaattu ulundu, Karamani
- **Telugu:** Alasandalu, Kaaraamanulu
- **Nepali:** Boodee

Season: September to October

Parts used: Seeds

Characteristics (Mithen and Kibblewhite 1993):

1. Growth Habit: It is an annual or perennial herbaceous plant with an upright growth habit.

2. Stem: The stems are slender, trailing, or climbing, and can reach varying heights depending on the variety.

3. Leaves: The leaves are alternate, trifoliate, and have long petioles. Each leaflet is ovate or lanceolate in shape and has a pointed tip. The leaf margins may be smooth or slightly toothed.

4. Inflorescence: The flowers are borne in axillary racemes or clusters.

5. Flowers: The flowers are typically yellow, but they can also be white, purple, or reddish in some cultivars. The flowers are papilionaceous, meaning they have a butterfly-like shape with five petals. The upper petal is known as the standard, the two side petals are called wings, and the two lower petals are fused to form a boat-shaped structure known as the keel (Fig. **24.16**).

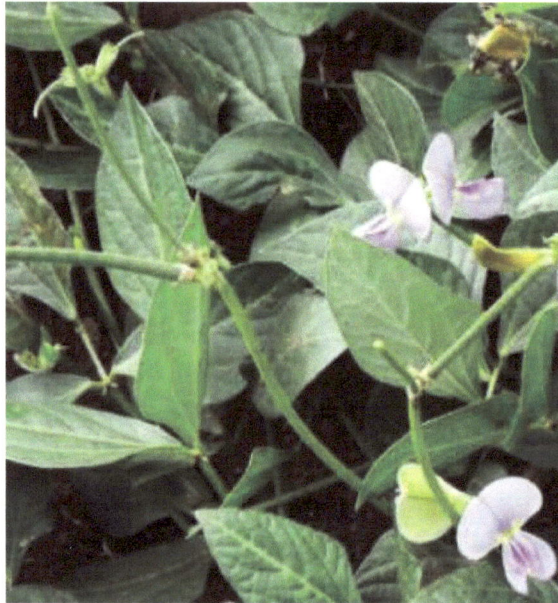

Fig. (24.16). *V. unguiculata* **Flowering twig with seeds** (PC: Apurva Shankar Chonde).

6. Pods: The fruit is a pod that varies in length, shape, and color depending on the cultivar. It can be straight or curved, and the surface may be smooth or slightly ribbed. The pods contain several seeds.

7. Seeds: The seeds are typically kidney-shaped or oval, and their color can range from white, cream, beige, brown, or black, depending on the cultivar. The seeds have a prominent "eye" or dark spot on one side, giving rise to the name "black-eyed pea" for some varieties (Fig. **24.16**).

Distribution:

Africa- Nigeria, Ghana, Senegal, Mali, Cameroon, and Sudan. Asia- India, Bangladesh, Pakistan, Myanmar, Thailand, Vietnam, and Indonesia, Americas- USA, Brazil, Mexico, Colombia, Venezuela, and the Caribbean islands. Europe- Italy, Greece, and Spain.

Propagation: Seed

Chemical constituents:

Proteins and amino acids, carbohydrates, vitamins, minerals, flavonoids, phenolic compounds, phytosterols, and triterpenoids.

Recipe:

Ingredients: Moong dal, leaves, mustard, cumin, green or red chilies, turmeric powder, hing, and garlic.

Method: Cook the dish using soaked moong dal. Use the leaves, particularly the central ones, from mature leaves. Wash and chop the leaves. Heat oil in a pan and add a small amount of mustard and cumin seeds. Then, add hing (or garlic, depending on your preference) and green or red chilies. Add the soaked dal (make sure to drain it first) followed by the leaves, turmeric, and salt to taste. Cook until the leaves become tender.

OR

Peel pods and chop them. Heat oil in a pan and fry asafoetida, mustard seeds, curry leaves, crushed garlic cloves, and finely chopped onion. When the onion turns slightly yellow, add the chopped pods, chili, turmeric powder, and salt to taste. Add water, cover the pan, and cook over low heat. Once the vegetables are cooked, add wet coconut on top.

Uses:

1. Food Source: The seeds are consumed as a nutritious food source in many parts of the world. They can be cooked, boiled, or ground into flour to make a variety of dishes such as soups, stews, curries, salads, and side dishes (Ouédraogo, 2003; Dugje *et al.*, 2009).

2. Dietary Fiber: Cowpea seeds are rich in dietary fiber, which aids in digestion and promotes a healthy digestive system. Including cowpea in the diet can help regulate bowel movements and prevent constipation (Dugje *et al.*, 2009).

3. Protein Source: Cowpea seeds are a valuable source of plant-based protein, particularly in regions where meat consumption is limited. They provide essential amino acids and can be used as a protein supplement in vegetarian and vegan diets (Pungulani *et al.*, 2013).

4. Nutritional Value: Cowpea is rich in various nutrients, including vitamins (A, B, and C), minerals (iron, calcium, and potassium), and antioxidants. Regular consumption of cowpeas can contribute to a well-balanced and nutritious diet (Hall *et al.*, 2003; Diouf, 2011).

5. Livestock Feed: Cowpea foliage, pods, and leftover plant material after seed harvest can be used as animal feed. It provides a good source of nutrition for livestock, such as cattle, goats, and sheep (Ouédraogo, 2003; Dugje *et al.*, 2009).

6. Green Manure: Cowpea plants have the ability to fix nitrogen from the air and enrich the soil with nitrogen, making them suitable for use as a green manure crop. They can be grown in rotation with other crops to improve soil fertility and enhance agricultural productivity (Asiwe *et al.*, 2009).

7. Soil Erosion Control: Cowpea plants with their deep root systems help control soil erosion. They are often planted as cover crops or intercropped with other crops to prevent soil degradation and enhance soil health (Asiwe *et al.*, 2009; Dugje *et al.*, 2008).

Vigna vexillata (L.) A. Rich.

Botanical name: *Vigna vexillata*

Family: Fabaceae

Local name: Halunda

- **Common name:** Zombi pea, Wild Mung, Wild cowpea
- **Hindi:** Junglee Mung, Suryapani, Machali
- **Marathi:** Halunda
- **Malayalam:** Kattupayar
- **Sanskrit:** Mudgaparni

Season: July to October

Parts used: All parts

Characteristics (Tripathi *et al.*, 2021):

1. Growth Habit: *V. vexillata* is a climbing or trailing perennial herb.

2. Stem: The stems are slender, wiry, and often covered with fine hairs.

3. Leaves: The leaves are alternate and trifoliate, consisting of three leaflets. The leaflets are elliptic to ovate in shape and have pointed tips. They are typically hairy on both surfaces.

4. Leaf Petioles: The leaf petioles are long and often have sparse hairs.

5. Flowers: The flowers are borne in axillary racemes or clusters. They are usually yellow or yellowish-white in color.

6. Inflorescence: The inflorescence is axillary, with several flowers arranged in a raceme.

7. Calyx: The calyx is tubular and has five lobes.

8. Corolla: The corolla is papilionaceous, meaning it has a large upper petal called the standard, two lateral petals called wings, and two lower petals fused to form the keel.

9. Fruit: The fruit is a legume, typically cylindrical or slightly curved. It contains several seeds (Fig. **24.17**).

Fig. (24.17). Fruits of *V. vexillata* (PC: Vasant Kale).

10. Seeds: The seeds are small, oval-shaped, and often have a mottled or speckled appearance. They vary in color, ranging from brown to black.

Distribution:

China, India, Ivory Coast, Jamaica, Japan, Kenya, Korea, Sri Lanka, Sudan, Suriname, Swaziland, Tanzania, and Togo.

Propagation: Seed

Chemical constituents:

Sterol, triterpenes, isoflavone, flavonoids, saponins, amino acids, fatty acids, phenolic acids, and alkaloids (Leu *et al.*, 2012).

Recipe:

Ingredients: Younger pods, finely chopped onion, 5-6 garlic cloves, salt, green chilies, curry leaves, Asafoetida, grated coconut, mustard seeds, turmeric, *etc.*

Method: Peel pods and chop them. Heat oil in a pan and fry asafoetida, mustard seeds, curry leaves, crushed garlic cloves, and finely chopped onion. When the onion turns slightly yellow, add the chopped pods, chili, turmeric powder, and salt to taste. Add water, cover the pan, and cook over low heat. Once the vegetables are cooked, add wet coconut on top.

Uses:

1. Food source: The seeds of *V. vexillata* are edible and can be used as a food source. They can be cooked and used in various dishes, including soups, stews, curries, and salads (Tripathi *et al.*, 2021).

2. Traditional medicine: In traditional medicine systems, various parts of *V. vexillata*, such as leaves, stems, and roots, are used for their medicinal properties. They have antioxidant, anti-inflammatory, and antimicrobial properties, and are used to treat various ailments (Doblado *et al.*, 2005).

3. Livestock feed: The foliage and seeds of *V. vexillata* can be used as fodder for livestock, providing a source of nutrition for animals (Tripathi *et al.*, 2021).

4. Green manure: *V. vexillata* is also used as a green manure crop. It is grown and then plowed back into the soil to improve soil fertility and nutrient content (Tripathi *et al.*, 2021).

CHAPTER 2

Wild Vegetables of the Family Lythraceae

INTRODUCTION

The Lythraceae family, commonly known as the loosestrife family, comprises approximately 620 species within 32 genera (Xu and Deng 2017). The members are rich in antioxidant compounds such as flavonoids, triterpenoids, and phenolic compounds.

Ammannia baccifera L.

Botanical name: *Ammannia baccifera* L.

Family: Lythraceae

Local name: Dhan bhaji, Aginbuti, Bharajambhula

Vernacular name:

- **Assamese:** Agnigarbha
- **Bengali:** Banmarich
- **English:** Blistering Ammania
- **Hindi:** Aginbuti, Jungli Mehendi
- **Kannada:** Kaadugida
- **Malayalam:** Kallur Vanchi
- **Sanskrit:** Agnigarbha, Kshetrabhusha, Pasanabheda
- **Tamil:** Kal-l-uruvi
- **Telugu:** Agnivendapaku

Season: Flowering and Fruiting: August to January

Parts used: Leaves

Characteristics:

1. Habit: It is an annual or perennial herbaceous plant (Fig. **25.1**).

Fig. (25.1). Whole plant of *A. baccifera* (PC: Vishnu Birajdar).

2. Stem: The stem is erect or decumbent, branched, and often reddish in color. It can grow up to 50 centimeters in height (Flora of North America Editorial Committee, 1997).

3. Leaves: The leaves are opposite, simple, lanceolate to linear-lanceolate in shape, and usually sessile or short-petiolate. They are green or reddish and may have toothed margins.

4. Flowers: The flowers are small, pink, or reddish in color, and occur in clusters in the leaf axils or at the stem tips. Each flower has four or five petals and numerous stamens.

5. The fruits are small, globose to ellipsoid capsules, which contain numerous tiny seeds (Wiart, 2006; Weeds Identification and Knowledge in the Western Indian Ocean).

Distribution:

It has a wide distribution and is native to tropical and subtropical regions. It is found in various countries across Africa, Asia, Australia, and the Americas.

Propagation: Seed

Chemical constituents:

It is rich in flavonoids, triterpenoids, and phenolic compounds. It consists of hentriacontine, dotriacontanol, betulinic acid, lupeol, ellagic acid, quercetin, and lawsone (Joshi, 2007).

Recipe:

Ingredients: Younger leaves, finely chopped onion, 5-6 garlic cloves, salt, green chilies, curry leaves, Asafoetida, grated coconut, mustard seeds, turmeric, Jaggery, gram dal, *etc.*

Method: Peel the stalks of the leaves, wash one judy leaf, and chop finely. Heat oil in the pan and fry asafoetida, mustard seed, curry leaves, crushed garlic cloves, and finely chopped onion. When the onion turns slightly yellow, add the chopped vegetables, chili, turmeric powder, soaked dal, jaggery and salt to taste; lightly beat water and cover, and cook over low flame. After the vegetables have finished cooking, sprinkle grated wet coconut over them.

Uses:

1. Edible Leaves: The leaves are consumed as a leafy vegetable and used in traditional culinary preparations (Nirmal *et al.*, 2011).

2. Anti-inflammatory Activity: Plant extract have shown potential in reducing inflammation and related conditions (Nirmal *et al.*, 2011).

3. Antioxidant Activity: The extracts have exhibited antioxidant activity, which can help neutralize harmful free radicals in the body and protect against oxidative stress-related damage (Vijayakumar *et al.*, 2012).

4. Wound Healing: Traditional medicinal uses include its application on wounds to promote healing. The plant extract has shown wound-healing properties in animal studies (Rajasekaran *et al.*, 2012)

5. Skin: The fresh, bruised leaves have been used in skin diseases as a rubefacient and as an external remedy for ringworm and parasitic skin affection (Nirmal *et al.*, 2011).

CHAPTER 3

Wild Vegetables of the Family Malvaceae

INTRODUCTION

The Malvaceae, commonly referred to as the mallow or hibiscus family, is a diverse group of flowering plants that includes over 4,225 species across 244 genera (Erarslan and Koçyiğit 2019). Some members are a good source of proteins, vitamin C, and minerals like potassium, phosphorous, magnesium, calcium, sodium, iron, *etc.*

Abelmoschus ficulneus (L.) Wight & Arn. ex Wight

Botanical name: *Abelmoschus ficulneus*

Family: Malvaceae

Local name: Ran Bhendi

Vernacular name:

- **Common name:** White Wild Musk Mallow, Native rosella
- **Hindi:** Jangli bhindi
- **Marathi:** Ran bhendi
- **Tamil:** Kattu-vendai
- **Telugu:** Nella benda, Parupubenda
- **Kannada:** Sanna bende

Season: flowering & fruiting: November-February

Parts used: Leaves, fruit, seeds

Characteristics (Native Rosella, 2008; White Wild Musk Mallow, 2010)

1. Plant Habit: It is a perennial herbaceous plant that can grow up to 2 meters in height. It may have a shrub-like growth form.

2. Leaves: The leaves are alternate and palmately lobed or divided. Each leaf typically has 3 to 7 lobes or divisions. The lobes are usually elongated and lanceolate in shape.

3. Flowers: The flowers are typically white in color, although they may have slight variations in shades. The flowers are large and showy, with a characteristic hibiscus-like appearance. They have five petals and a prominent central column of stamens (Fig. **26.1**).

Fig. (26.1). **Flower and fruits of** *A. ficulneus* (PC: Apurva Shankar Chonde).

4. Fruit: After flowering, the plant produces fruit capsules. These capsules are generally round or ovoid in shape and contain numerous seeds (Fig. **26.1**).

5. Seed Characteristics: The seeds are small and often flattened or disc-shaped. They may have a dark coloration.

Distribution:

Asia: India, Malaysia, Pakistan, Sri Lanka; Africa: Madagascar; Australasia.

Propagation: Seed

Chemical constituents:

Leaves have beta-sitosterol and beta-D glucoside. Flowers have anthocyanins. Petals have beta-sitosterol, its glucoside, and the flavonoid myricetin. The seeds have an essential oil containing farnesol, palmitic acid, acetic acid, furfural, lactose, and amaretto acid (Sharma, 1993).

Recipe:

Ingredients: Bhendi, onion, garlic cloves, green chilies, grated coconut, turmeric powder, asafoetida, salt, oil.

Method: Heat oil in a pan, add onion, and sauté 1 minute, now add garlic, chilies, asafoetida, turmeric, Bhendi, and salt mix well. Take a lid, and cook the Bhendi on low flame, keep on checking in between and stir continuously, so that Bhendi does not stick to the bottom of the pan. When the Bhendi is cooked, add coconut and mix well. It can be served with roti or chapati.

Uses:

1. The whole plant is utilized in traditional medicine to treat conditions such as sprains, bronchitis, and toothache (Mohite *et al.*, 2019; Dashputre *et al.*, 2021).

2. It exhibits various pharmacological activities, including its potential in treating jaundice and other serious illnesses (Florence, 2019).

3. White Wild Musk Mallow is known to have therapeutic effects on upset stomach and diarrhea. Drinking a morning tea prepared from its leaves can provide relief (Sharma, 1993).

4. The Native Rosella species is used to treat scorpion bites by applying crushed roots directly to the affected area and consuming juice made from the crushed roots (Sharma, 1993).

5. Native Rosella is beneficial for stomach health as it strengthens the stomach muscles and promotes proper gastric function (Sharma, 1993).

6. It aids in liver function by enhancing its activity. People with jaundice can experience faster recovery by consuming crushed roots of the plant with water, twice a day (Sharma, 1993).

7. Consuming a cup of decoction made from the leaves (or roots for a stronger effect) in the morning can help open airways and assist in managing asthma (Sharma, 1993).

8. It has a cooling effect on the body, inducing a sense of calmness when a decoction of its leaves and roots is consumed (Sharma, 1993).

9. The tonic prepared from the roots of the plant aids in improving heart health. It helps lower cholesterol levels and prevents arterial and blood vessel scaling (Sharma, 1993).

10. The seed emulsion of *A. ficulneus* possesses antispasmodic properties, particularly effective in the digestive system. Chewing the seeds acts as a nervine and stomachic, while also freshening breath. The seeds are valued for their diuretic, demulcent, and stomachic properties. They also showed stimulant, antiseptic, cooling, tonic, carminative, and aphrodisiac effects (Sharma, 1993).

11. A paste made from the bark is topically applied to cuts, wounds, and sprains. The essential oil of *A. ficulneus* is used in aromatherapy for depression and anxiety treatment. The leaf decoction has been effective against intestinal complaints and checks vomiting. The tincture of leaf powder is applied for skin diseases (Grieve, 1984).

12. Externally, essential oil is applied to alleviate cramps, improve circulation, and relieve joint pain (Venugopalan, 2017; Thamizhselvam *et al.*, 2020).

Bombax ceiba L.

Botanical name: *Bombax ceiba*

Family: Malvaceae

Local name: Saanvar, Red Cotton Tree

Vernacular name:

- **Common name:** Silk Cotton Tree, Kapok Tree
- **Hindi:** Shalmali, Semal
- **Manipuri:** Tera
- **Assamese:** Dumboil
- **Kannada:** Kempu booraga, Kempu booruga, Elava
- **Tamil:** Sittan, Sanmali
- **Malayalam:** Unnamurika
- **Nepali:** Simal
- **Mizo:** Phunchawng
- **Sanskrit:** Shaalmali, Sthiraayu

Season: Leaf Fall: December- March; Flowering: January- February; Fruiting: March-May.

Parts used: Fruits, heartwood, stem bark, gum, and root.

Characteristics (Nicolson, 1979):

1. Tree Size: It is a large, deciduous tree that can reach heights of 20 to 40 meters or more.

2. Trunk: The trunk is typically straight and cylindrical, with a diameter ranging from 1 to 2 meters. It is often characterized by large buttresses at the base for stability.

3. Bark: The bark is grayish-brown and smooth when young, but it becomes rough and fissured with age.

4. Leaves: The leaves are compound, palmately lobed, and alternate. Each leaf is composed of 5 to 9 leaflets, which are ovate or lanceolate in shape. The leaflets have serrated edges and are usually 10 to 20 centimeters long.

5. Flowers: It produces large, showy flowers that are typically red or pink in color. The flowers are cup-shaped and have five petals. They are borne in clusters at the ends of branches (Fig. **26.2**).

Fig. (**26.2**). *B. ceiba* **Flower** (PC: Tukaram Kanoja).

6. Fruits: The fruits are large, woody capsules that contain numerous seeds embedded in cotton-like fibers. When the fruits mature, they split open, releasing the seeds that are dispersed by wind.

Distribution:

E. Asia - southern China, Indian subcontinent, Myanmar, Thailand, Laos, Vietnam, Malaysia, Indonesia, Philippines to Papua New Guinea and Australia.

Propagation: Seed, cuttings of half-ripe wood

Chemical constituents:

Flavonoids- quercetin, kaempferol, and their glycosides, triterpenoids- beta-sitosterol, lupeol, and their derivatives, phenolic compounds- gallic acid, protocatechuic acid, and caffeic acid derivatives, steroids- stigmasterol and beta-sitosterol, alkaloids- bombaxine (Wang *et al.*, 2014).

Recipe:

The flower buds and the calyx of not fully opened flowers are eaten cooked as a vegetable.

Uses:

1. Edible Parts: The young leaves, flowers, and young fruits are reported to be edible in some cultures. They are used in traditional culinary practices, such as being cooked and consumed as vegetables or added to soups and stews (Zahan *et al.*, 2013).

2. Dietary Fiber: It may contain dietary fiber, which is important for digestive health. Including fiber-rich foods in the diet can help promote regular bowel movements and support overall gastrointestinal health (Zahan *et al.*, 2013).

3. Traditional Medicine: Various parts of *B. ceiba*, including the bark, leaves, flowers, and seeds, have been used in traditional medicine for their potential therapeutic properties. They have been used to treat conditions such as diarrhea, dysentery, fever, skin diseases, respiratory ailments, and inflammation (Gupta *et al.*, 2022).

4. Antimicrobial Activity: Extracts from different parts of *B. ceiba* have shown antimicrobial activity against certain bacteria and fungi. They have been used in traditional remedies to help combat infections (Gupta *et al.*, 2022).

5. Wound Healing: The latex obtained from the stem bark of *B. ceiba* has been traditionally applied topically to wounds to promote healing and provide relief (Gupta *et al.*, 2022).

6. Anti-inflammatory Properties: Some studies have reported the anti-inflammatory effects of *B. ceiba* extracts. They have been used in traditional medicine to alleviate inflammation and related conditions (Zahan *et al.*, 2013; Tundis *et al.*, 2014).

7. Antioxidant Activity: Extracts from *B. ceiba* have shown antioxidant properties, which can help protect against oxidative stress and associated damage in the body (Zahan *et al.*, 2013; Tundis *et al.*, 2014).

Corchorus olitorius L.

Botanical name: *Corchorus olitorius*

Family: Malvaceae

Local name: Chuch

Vernacular name:

- **Common name:** Nalta Jute, Jew's Mallow, Tossa jute
- **Hindi:** Pat, Pat-sag, Mithapat
- **Manipuri:** Limon
- **Marathi:** Motichhunchh, Banpat
- **Tamil:** Punaku, Peratti
- **Telugu:** Parinta
- **Bengali:** Bhungipat
- **Oriya:** Kaunria
- **Sanskrit:** Mahachanchu

Season: January to December

Parts used: Leaves

Characteristics (Gupta *et al.*, 2023):

1. Plant Height: *C. olitorius* is an herbaceous annual plant that can grow up to 2-4 meters in height.

2. Leaves: The leaves of *C. olitorius* are alternate, simple, and have a lanceolate or ovate shape. They are typically dark green in color and have a slightly toothed margin. The leaves are arranged spirally along the stems.

3. Flowers: The flowers of *C. olitorius* are small and yellow in color. They are borne on long stalks and have five petals. The flowers are typically found in clusters at the leaf axils (Fig. **26.3**).

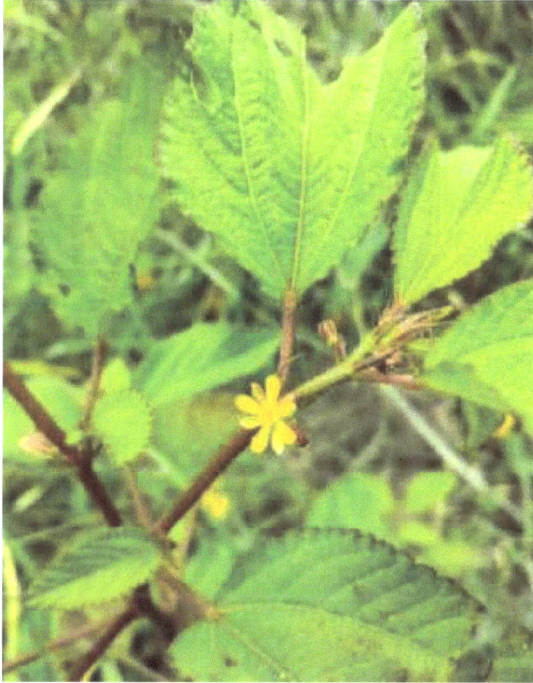

Fig. (26.3). Flowering twig of C. *olitorius* (PC: Vijesh Kumar).

4. Fruits: After flowering, *C. olitorius* produces capsule-like fruits that are cylindrical or oblong in shape. The fruits contain several seeds.

5. Stems: The stems of *C. olitorius* are slender, branched, and often reddish or purplish in color. They have a fibrous texture.

6. Roots: *C. olitorius* has a fibrous root system that helps anchor the plant in the soil.

Distribution:

Africa- Nigeria, Ghana, Sudan, Ethiopia, Kenya, Tanzania, Uganda, and South Africa.

Asia- India, Bangladesh, Pakistan, Sri Lanka, Myanmar, Thailand, Malaysia, Indonesia, and the Philippines. Europe- Greece, Cyprus, and Turkey. Americas- Brazil, Venezuela, Colombia, and the Caribbean islands.

Propagation: Seed

Chemical constituents:

Flavonoids- quercetin, kaempferol, and apigenin, Phenolic compounds- caffeic acid, chlorogenic acid, and protocatechuic acid, vitamins- vitamin C, vitamin A (beta-carotene), and vitamin E, Minerals- calcium, iron, magnesium, and potassium, omega-3 fatty acids, Polysaccharides- galactomannans.

Recipe:

Ingredients: Younger leaves, finely chopped onion, 5-6 garlic cloves, salt, green chilies, curry leaves, asafoetida, grated coconut, mustard seeds, turmeric, gram dal, *etc.*

Method: Wash the Judy leaves and finely chop them. Heat oil in a pan and add asafoetida, mustard seeds, curry leaves, crushed garlic cloves, and finely chopped onion. Once the onion turns slightly yellow, add the chopped vegetables, chili, turmeric powder, soaked dal, and salt to taste. Lightly beat water and cover the pan. Cook the mixture over low heat until the vegetables are tender. Finally, garnish with grated wet coconut.

Uses:

1. Culinary: The leaves of *C. olitorius* are edible and commonly used as a vegetable in various cuisines. They are added to soups, stews, curries, and stir-fries. The leaves have a mild and slightly nutty flavor (Ola *et al.*, 2009).

2. Nutritional Benefits: *C. olitorius* is highly nutritious and rich in vitamins, minerals, and dietary fiber (Jaarsveld *et al.*, 2014; Sarker *et al.*, 2015; Sivakumar *et al.*, 2018). It is a good source of vitamins A, C, and E, iron, calcium, and antioxidants. Consuming jute leaves can contribute to a healthy and balanced diet.

3. Medicinal: In traditional medicine, *C. olitorius* is used for its various medicinal properties. It shows anti-inflammatory, antioxidant, and antimicrobial properties. The leaves are used in the treatment of various ailments such as cough, fever, and digestive disorders (Sivakumar *et al.*, 2018; Maseko *et al.*, 2018; Conti *et al.*, 2019).

4. Fiber Production: *C. olitorius* is primarily cultivated for its long, strong fibers that are used in the production of jute textiles and ropes. The plant is widely grown for its commercial value in the textile industry (İşeri *et al.*, 2013).

Corchorus trilocularis L.

Botanical name: *Corchorus trilocularis*

Family: Malvaceae

Local name: Kadu Choonch

Vernacular name:

- **Common name:** Wild Jute, African jute, Three locule corchorus
- **Hindi:** Kadukosta, Kadvapat, Hardikaket
- **Kannada:** Ennepundi gida, Kadu Choonchali gida
- **Marathi:** Kadu Choonch
- **Oriya:** Jangali jiraa
- **Sanskrit:** Chanchu, Dirghachanchu, Kaunti, Raja-jira
- **Telugu:** Bankitutturu

Season: September

Parts used: Leaves

Characteristics:

1. Plant Size: *C. trilocularis* is a perennial shrub that can grow up to 1-2 meters in height. It has a branching habit and a bushy appearance.

2. Leaves: The leaves of *C. trilocularis* are alternate, simple, and palmately veined. They are dark green in color and have a lobed or palmate shape. The leaves are broader towards the base and taper to a point at the tip.

3. Leaf Margins: The leaf margins of *C. trilocularis* are usually serrated or toothed, with irregularly spaced teeth along the edges.

4. Flowers: The flowers of *C. trilocularis* are small and yellow in color. They are borne on short stalks and typically have five petals. The flowers are arranged in clusters or cymes at the ends of the branches (Fig. **26.4**).

Fig. (26.4). Flowering twig of *C. trilocularis* (PC: Rajesh Ghumatkar).

5. Fruits: The fruits of *C. trilocularis* are cylindrical capsules that are green when young and turn brown as they mature. The capsules contain numerous seeds.

6. Stems: The stems of *C. trilocularis* are slender, green, and slightly woody at the base. They have a smooth texture.

Distribution:

West Africa- Nigeria, Ghana, Cameroon, Ivory Coast, Senegal, and Burkina Faso; East Africa- Ethiopia, Kenya, Tanzania, Uganda, and Sudan.; Central Africa- Congo, Democratic Republic of Congo, and Central African Republic; Southern Africa- Angola, Mozambique, Malawi, and Zambia.

Propagation: Seed

Chemical constituents: Flavonoids- quercetin, kaempferol, and rutin. Phenolic compounds- caffeic acid, chlorogenic acid, and ferulic acid. Triterpenoids- betulin, betulinic acid, and oleanolic acid, Sterols- β-sitosterol and stigmasterol.

Recipe:

Ingredients: Younger leaves, finely chopped onion, 5-6 garlic cloves, salt, green chilies, curry leaves, asafoetida, grated coconut, mustard seeds, turmeric, gram dal, *etc.*

Method: Peel the stalks of the leaves and wash the Judy leaves. Finely chop the leaves. Heat oil in a pan and add asafoetida, mustard seeds, curry leaves, crushed garlic cloves, and finely chopped onion. Once the onion turns slightly yellow, add the chopped vegetables, chili, and turmeric powder. Also, add soaked dal and salt to taste. Lightly beat water and cover the pan. Cook the mixture over low heat until the vegetables are tender. Once the vegetables are cooked, sprinkle wet coconut on top.

Uses:

1. Culinary: The leaves of *C. trilocularis* are edible and can be cooked and consumed as a leafy vegetable. They are often used in soups, stews, and stir-fries in various cuisines (Dhanalakshmi and Manavalan 2014).

2. Medicinal: In traditional medicine, *C. trilocularis* is used for its medicinal properties. It also showed diuretic, anti-inflammatory, and antioxidant effects. The plant is used to treat various ailments such as fever, cough, asthma, and gastrointestinal disorders (Ahirrao *et al.*, 2009).

Hibiscus sabdariffa **L.**

Botanical name: *Hibiscus sabdariffa*

Family: Malvaceae

Local name: Roselle, Laal-ambaari, tambdi-ambadi

Vernacular name:

- **Common name:** Roselle, Hibiscus, Jamaica sorrel, red sorrel
- **Angami:** Gakhro
- **Hindi:** Lal Ambari, Patwa
- **Manipuri:** Silo-sougree
- **Marathi:** Laal-ambaari, Tambdi-ambadi
- **Tamil:** Simaikkasuru, Sivappukkasuru, Shimai-kashuruk-kirai
- **Malayalam:** Polechi, Puli-cheera
- **Telugu:** Erragomgura, Erragonkaya, Ettagomgura
- **Kannada:** Kempupundrike, Pulichakire, Pundibija
- **Bengali:** Chukar
- **Mizo:** Lekhar-anthur
- **Assamese:** Chukiar, Tengamora
- **Sanskrit:** Ambasthaki

Season: Flowering: October-November.

Parts used: Calyx, seed, leaves.

Characteristics (Morton, 1987, Ross, 2003):

1. Habit: *H. sabdariffa* is an annual or perennial herbaceous plant that can reach a height of 2 to 3 meters (6 to 10 feet). It has an upright and bushy growth habit (Fig. **26.5**).

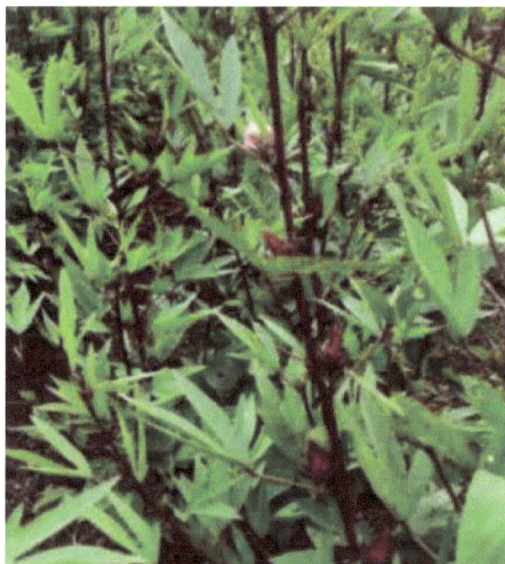

Fig. (26.5). Whole plant of *H. sabdariffa* (PC: Prajkta Nandgaonkar).

2. Leaves: The leaves of *H. sabdariffa* are alternate, simple, and palmately lobed. Each leaf typically has three to five lobes, with serrated or toothed margins. The leaves are dark green in color and have a rough texture.

3. Flowers: The flowers of *H. sabdariffa* are large and showy, with a diameter of about 8 to 10 centimeters (3 to 4 inches). They are typically bright red or deep pink, although some cultivars may have white or pale yellow flowers. The petals are broad and overlapping, giving the flower a trumpet-like appearance.

4. Calyx: One of the key identification features of *H. sabdariffa* is its prominent and enlarged calyx, which surrounds the base of the flower. The calyx is fleshy, and reddish, and consists of several sepals that are fused together. When the flower matures, the calyx expands and forms a cup-like structure that encloses the seed pod.

5. Seed Pod: After the flowers fade, the calyx of *H. sabdariffa* enlarges further, forming a distinctive seed pod or fruit. The fruit is round or conical in shape and becomes woody when mature. It is typically about 2 to 3 centimeters (0.8 to 1.2 inches) in diameter and has a reddish-brown color.

Distribution: Africa- Nigeria, Senegal, Sudan, Mali, Ghana, and others, Asia-India, Thailand, China, Malaysia, Indonesia, Sri Lanka, Vietnam, and the Philippines, Americas- Mexico, the Caribbean islands, Central America, and South America. Countries such as Jamaica, Haiti, Mexico, Brazil, and Colombia.

Propagation: Seed, Cuttings of half-ripe wood

Chemical constituents:

Anthocyanins, Organic acids- citric acid, malic acid, tartaric acid, and hibiscus acid, Vitamin C, polyphenols- flavonols, flavan-3-ols, and proanthocyanidins, polysaccharides, volatile compounds, *etc.*

Recipe:

Ingredients: buds, garlic, onion, oil, chili powder, salt, *etc.*

Method: Wash and chop the flower buds. Heat oil in a pan and fry crushed garlic cloves, followed by chopped vegetables, red chilies, and salt. Allow the vegetables to steam thoroughly. Stir the vegetables constantly to ensure even cooking and to prevent sticking. Stirring will help loosen the vegetables. Cook the vegetables over low flame.

Uses:

1. Culinary: The calyxes (flower sepals) of *H. sabdariffa* are used in culinary preparations. They are often dried and used to make herbal teas, infusions, and beverages. The tart flavor of Roselle infusions is enjoyed as a refreshing drink, either hot or cold. The calyxes can also be used in the preparation of jams, jellies, sauces, and desserts (Bako *et al.*, 2009, Bolade *et al.*, 2009).

2. Traditional Medicine: *H. sabdariffa* has a long history of use in traditional medicine systems. It has various health benefits, including potential anti-inflammatory, antioxidant, and diuretic properties. In traditional medicine, Roselle has been used to manage conditions such as high blood pressure, liver disorders, urinary tract infections, and digestive issues (Morton, 1987; Shalgum *et al.*, 2018; Islam, 2019; Izquierdo *et al.*, 2020; Al-Yousef *et al.*, 2020).

3. Antioxidant and Anti-inflammatory Effects: Studies have suggested that *H. sabdariffa* exhibits antioxidant and anti-inflammatory properties due to its rich content of anthocyanins and polyphenolic compounds. These properties may help protect against oxidative stress and inflammation-related diseases (Gurrola-Díaz *et al.*, 2010; Hassan *et al.*, 2016; Shalgum *et al.*, 2018; Islam, 2019; Izquierdo *et al.*, 2020; Al-Yousef *et al.*, 2020).

4. Cardiovascular Health: *H. sabdariffa* has been studied for its potential cardiovascular benefits. Research suggests that consuming Roselle infusions may help lower blood pressure and improve lipid profiles by reducing LDL cholesterol and triglyceride levels (Talal Al-Malki *et al.*, 2021).

Urena lobata L.

Botanical name: *Urena lobata*

Family: Malvaceae

Local name: vanbhendi

Vernacular name:

- **Assamese:** Shonboroluwa, Sokamora
- **Bengali:** Bana okara
- **Gujarati:** Vagadau bhindo
- **Hindi:** Bachita, Lapetua, Vilaiti san
- **Kachchhi:** San bhindo
- **Kannada:** Dodda bende, Kaadu thutthi, Otte
- **Konkani:** Tupkadi
- **Malayalam:** Oorppanam, Uram, Uthiram
- **Manipuri:** Sampakpi
- **Marathi:** Ran tupkuda, Vanbhendi
- **Mizo:** Se-hnap
- **Nepali:** Bariyaar, Bherejhaar, Dalle kuro, Katahare kuro, Naalukuro
- **Odia:** Jatajatia
- **Sanskrit:** Atibala, Bala, Nagabala
- **Tamil:** Ottu-t-tutti
- **Telugu:** Nalla benda, Padanikaada, Pedda benda, Piliyamankena, Vana benda
- **Tulu:** Boldu urki
- **Tangkhul:** Phanang

Season: July to September

Parts used: Leaves, roots, stem, and bark

Characteristics (Austin, 1999):

1. Habitat: *U. lobata* is commonly found in tropical and subtropical regions. It thrives in moist, well-drained soils and is often found in disturbed areas, such as fields, roadsides, and waste areas.

2. Plant Type: *U. lobata* is an herbaceous perennial plant.

3. Height: It can grow up to 1-2 meters in height.

4. Leaves: The leaves of *U. lobata* are palmately lobed, meaning they have multiple lobes radiating from a central point. The lobes are usually 3-5 in number and are lanceolate or ovate in shape. The leaves have serrated margins and are covered with fine hairs (pubescent).

5. Flowers: *U. lobata* produces small, bell-shaped flowers that are typically pink, purple, or white in color. The flowers are arranged in clusters at the tip of the branches (Fig. **26.6**).

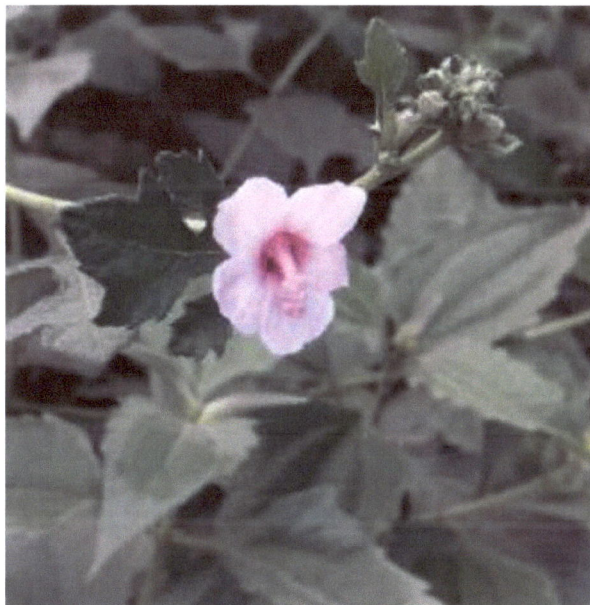

Fig. (26.6). Flower of *U. lobata* (PC: Apurva Shankar Chonde).

6. Fruits: The plant bears round or slightly flattened fruits that are covered with stiff, bristly hairs. The fruits are small and contain multiple seeds.

7. Stems: The stems of *U. lobata* are erect and branched. They are often reddish or purplish in color.

8. Hairs: The entire plant, including the leaves, stems, and fruits, is covered with fine, stiff hairs (pubescence). These hairs give the plant a rough and velvety texture.

Distribution:

It is found in the wild across the tropical and temperate zones of North and South America, as well as in Asia, including Indonesia, the Philippines, and Africa. It is also present in the Basin and Central Africa, with smaller cultivated areas in Brazil, India, and Madagascar.

Propagation: Seeds and stem cuttings.

Chemical constituents:

Alkaloids, flavonoids, glycosides, saponins, tannins, and terpenoids (Islam *et al.*, 2017).

Recipe:

Seeds - added to stews and porridges. Young leaves - cooked and used as a vegetable.

Uses:

1. Medicinal Purposes: *U. lobata* has been used in traditional medicine for its medicinal properties. It possesses diuretic, anti-inflammatory, and antimicrobial properties. Infusions or decoctions made from the leaves or roots of *U. lobata* are used to treat conditions such as urinary tract infections, digestive issues, and skin disorders (Islam *et al.*, 2021).

2. Fiber and Textiles: The bark of *U. lobata* contains strong and durable fibers. These fibers can be extracted and used for making ropes, cords, and coarse textiles. In some regions, the fiber is also used for making traditional crafts, baskets, and mats (Kengoh *et al.*, 2021).

CHAPTER 4

Wild Vegetables of the Family Menispermaceae

INTRODUCTION

The Menispermaceae family, commonly known as the moonseed family, is distinguished by a unique combination of features including woody climbers, palmate or pinnate venation, small and inconspicuous flowers, crescent-shaped seeds, *etc.* This family is represented by about 70 genera and 520 species (Chinh *et al.*, 2015). The members are rich in alkaloids.

Tinospora cordifolia (Willd.) Miers

Botanical name: *Tinospora cordifolia*

Family: Menispermaceae

Local name: Guduchi, Gulvel

Vernacular name:

- **English:** Gulancha/ Indian Tinospora
- **Sanskrit:** Guduchi, Madhuparni, Amrita, Chinnaruha, Vatsadaani, Tantrika, Kundalini & chakralakshanika.
- **Hindi:** Giloya, Guduchi
- **Bengali:** Gulancha
- **Telugu:** Thippateega
- **Tamil:** Shindilakodi
- **Gujarathi:** Galo
- **Kannada:** Amrita balli, Madhupa

Season: November to June

Parts used: Leaves, stem, root

Characteristics: (Upadhyay *et al.*, 2010):

1. Climbing Vine: *T. cordifolia* is a deciduous climbing vine that grows by twining around trees or other supports. It has a robust and woody stem with characteristic aerial roots that attach to surfaces for support.

2. Heart-shaped Leaves: The leaves of *T. cordifolia* are heart-shaped or cordate, hence the species name "cordifolia." The leaves are simple, alternate, and have a smooth or slightly wavy margin. They are typically green in color and can vary in size, with prominent veins running through them (Fig. **27.1**).

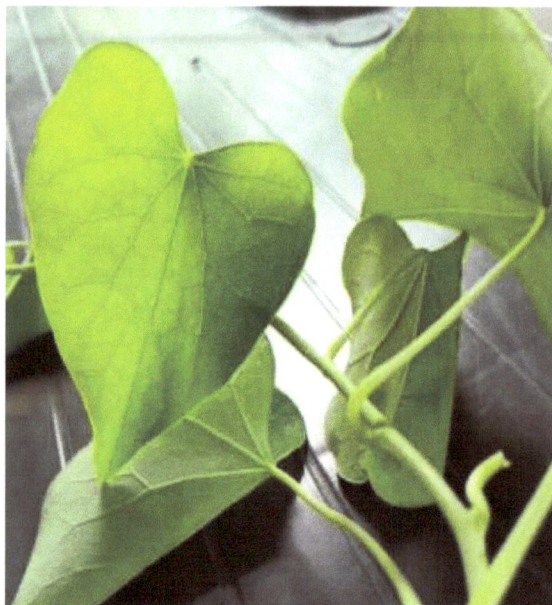

Fig. (27.1). Leaves of *T. cordifolia* (PC: Apurva Shankar Chonde).

3. Greenish-yellow Flowers: *T. cordifolia* produces small, unisexual flowers that are greenish-yellow in color. The flowers are typically arranged in axillary racemes or clusters.

4. Fleshy Stem: The stem of *T. cordifolia* is succulent and fleshy, with a pale green or brownish color. It is usually cylindrical, with nodes and internodes along its length.

5. Bitter Taste: When the stem of *T. cordifolia* is broken or crushed, it exudes a bitter taste.

Distribution: Bangladesh, Borneo, China, India, Indonesia, Malaysia, Myanmar, North Africa, Philippines, Sri Lanka, South Africa, Thailand, Vietnam and West Africa (Singh *et al.*, 2003).

Propagation: Seeds and cuttings.

Chemical constituents:

Alkaloids- Berberine, Columbin, Jatrorrhizine, Palmarin, Tetrahydropalmatine; Diterpenoid Lactones- Tinosporin, Tinosporide, Tinosporaside; Steroids-Gilosterol, Tinosporic acid, Tinosporol, Tinosporon, Beta-sitosterol; Arabinogalactan; Tinocordiside & Tinosporidine; Giloside; Syringin *etc.* (Upadhyay *et al.*, 2010; Saha and Ghosh, 2012).

Recipe:

Ingredients: leaves, chopped onion and garlic, oil, chili powder, salt, *etc.*

Method: Wash and chop the leaves. Fry onions and garlic in oil until they turn red. Then add the chopped vegetables, red chili powder, and salt. Sauté everything well. Steam and cook the vegetables.

Uses:

1. Immune System Support: *T. cordifolia* is believed to have immunomodulatory properties, meaning it helps regulate and support the immune system. It is traditionally used to enhance the body's natural defense mechanisms and strengthen immunity (Rege *et al.*, 1993; Nagarkatti *et al.*, 1994).

2. Fever and Respiratory Infections: *T. cordifolia* has been traditionally used to manage fever and respiratory infections, including common cold, flu, and cough. It may have antipyretic and antimicrobial properties (Sharma, 2001; Chunekar *et al.*, 2006).

3. Digestive Health: *T. cordifolia* is known for its digestive benefits. It may support healthy digestion, improve appetite, and help manage gastrointestinal disorders such as indigestion, bloating, and constipation (Sharma, 2001).

4. Antioxidant and Anti-inflammatory Effects: *T. cordifolia* contains various bioactive compounds that possess antioxidant and anti-inflammatory properties. These properties help combat oxidative stress, reduce inflammation, and may have a protective effect against chronic diseases (Stanely *et al.*, 2001; Wesley *et al.*, 2008).

5. Liver Health: *T. cordifolia* is used in traditional medicine to support liver health and promote detoxification. It has hepatoprotective properties and may help in managing liver disorders (Karkal *et al.*, 2007).

6. Diabetes Management: Some studies suggest that *T. cordifolia* may have antidiabetic properties and can help in managing blood sugar levels. It may enhance insulin secretion and improve glucose utilization in the body (Stanely *et al.*, 2001; Sharma, 2001; Prince *et al.*, 2004; Chunekar *et al.*, 2006).

7. Skin Health: *T. cordifolia* is traditionally used to promote healthy skin. It may have detoxifying and rejuvenating effects on the skin, helping in the management of various skin disorders (Asthana *et al.*, 2001; Sharma, 2001; Chunekar *et al.*, 2006).

CHAPTER 5

Wild Vegetables of the Family Moraceae

INTRODUCTION

The Moraceae family, often called the fig or mulberry family, has several key distinguishing characteristics such as milky sap, syconium fruit and unisexual flowers. The members are mostly tropical trees and consist of 37 genera and 1100 species (Francis 2004).

Ficus racemosa L.

Botanical name: *Ficus racemosa*

Family: Moraceae

Local name: Umber, Goolar

Vernacular name:

- **Hindi:** Goolar
- **Manipuri:** Heibong
- **Telugu:** Paidi
- **Sanskrit:** Udumbara
- **Malayalam:** Atti
- **Tamil:** Atti
- **Kannada:** Atti, Rumadi
- **Oriya:** Dimri
- **Nepali:** Gular, Dumri
- **Mizo:** Theichek

Season: Flowering: January- April; Fruiting: March-July

Parts used: All parts

Ganesh Chandrakant Nikalje, Apurva Chonde, Sudhakar Srivastava & Penna Suprasanna

Characteristics:

1. Size and Growth Habit: *F. racemosa* is a large deciduous tree that can reach heights of up to 30 meters (98 feet). It has a spreading canopy and a dense crown (Kirtikar *et al.*, 1975).

2. Trunk and Bark: The trunk of *F. racemosa* is typically straight and cylindrical, with a grayish-brown to dark brown bark. The bark may have vertical fissures or irregular patterns (Kirtikar *et al.*, 1975).

3. Leaves: The leaves of *F. racemosa* are alternate and simple, with an elliptical to obovate shape. They are leathery and glossy, measuring around 10-20 cm (4-8 inches) in length. The leaves have prominent veins and a smooth margin (Kirtikar *et al.*, 1975).

4. Fruits: One of the distinctive features of *F. racemosa* is its fig-like fruits, which grow in clusters along the branches and trunk. These figs are small, round to pear-shaped, and turn from green to yellow or orange when ripe. Each fig contains numerous tiny flowers (Kirtikar *et al.*, 1975) (Fig. **28.1**).

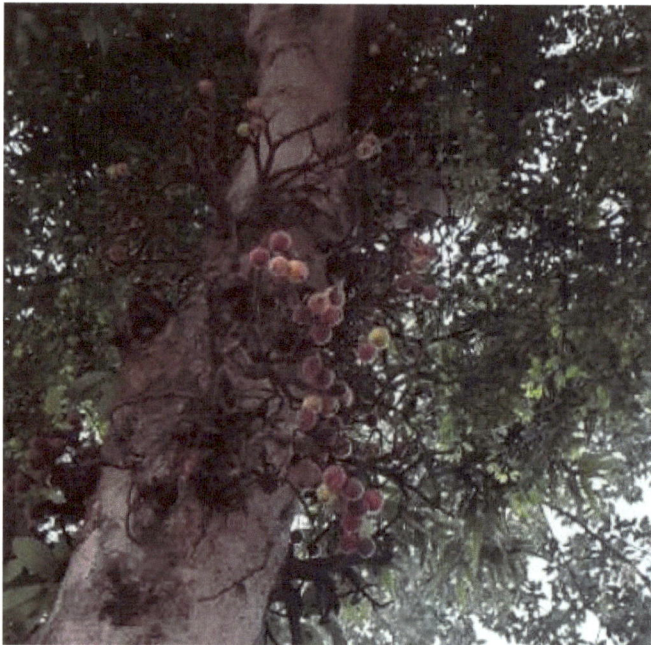

Fig. (28.1). Fruits of *F. racemosa* (PC: Apurva Shankar Chonde).

5. Roots: *F. racemosa* is known for its aerial roots, which are characteristic of many fig tree species. These roots grow down from the branches and trunk and

can form striking aerial root structures, especially in older trees (Kirtikar *et al.*, 1975).

Distribution:

China, India, Indonesia, Myanmar, Nepal, Papua New Guinea, Pakistan, Sri Lanka, Thailand, Vietnam, Australasia.

Propagation: Seed, Tip cuttings

Chemical constituents:

It contains tannin, wax, saponin gluanol acetate, β-sitosterol (A), leucocyanidin- 3 – O – β – D - glucopyrancoside, leucopelargonidin – 3 – O – β – D - glucopyranoside, leucopelargonidin – 3 – O – α – L - rhamnopyranoside, lupeol (C), ceryl behenate, lupeol acetate, α-amyrin acetate(B), leucoanthocyanidin and leucoanthocyanin from trunk bark lupeol, β-sitosterol and stigmasterol were isolated (Husain *et al.*, 1992).

Recipe:

Ingredients: Cleaned and chopped seed removed gooler, roasted chickpeas flour or sattu, minced ginger, garlic and green chilies, garam masala, coriander and mint greens, salt, ghee, *etc.*

Method: Boil the cleaned gooler in sufficient water until it becomes soft, then drain and allow it to cool. Mash it together with the other ingredients, excluding the ghee, to create a dough-like mixture. Shape the mixture into lime-sized balls and flatten them between your palms. Arrange the flattened balls to be shallow-fried in batches, using ghee or any oil of your choice. Serve them hot with green chutney or tamarind chutney.

Uses:

1. Medicinal: Various parts of *F. racemosa*, including the bark, leaves, and fruits, are used in traditional medicine for their potential health benefits. The plant possesses anti-inflammatory, analgesic, anti-diabetic, and antimicrobial properties. It is used to treat conditions such as diarrhea, dysentery, skin diseases, diabetes, and respiratory disorders. Tender fruits are astringent, stomachic, refrigerant, dry cough, loss of voice, diseases of kidney and spleen, astringent to bowel, styptic, tonic, useful in the treatment of leucorrhea, blood disorder, burning sensation, fatigue, and urinary discharges (Ahmed and Urooj 2010).

2. Food Source: The ripe fruits of *F. racemosa* are edible and can be consumed by humans and wildlife. In some regions, the sweet and juicy figs are eaten fresh or used in various culinary preparations. They can also be used in the preparation of jams, jellies, *etc.* (Ahmed and Urooj 2010).

3. Ecological Importance: *F. racemosa* plays a vital role in supporting biodiversity and ecosystem functioning. The fig fruits serve as a food source for numerous animals, including birds, bats, monkeys, and various insects. The trees also provide shelter and nesting sites for birds and other small animals (Ahmed and Urooj 2010).

4. Shade and Ornamental Tree: Due to its large size and spreading canopy, *F. racemosa* is often planted as a shade tree in parks, gardens, and along streets. The dense foliage provides shade and helps to cool the surrounding environment. Additionally, the tree's attractive leaves and figs make it a popular choice for ornamental purposes.

Ficus religiosa L.

Botanical name: *Ficus religiosa*

Family: Moraceae

Local name: Pippala tree, Pipal

Vernacular name:

- **Assamese:** Anhot
- **Bengali:** Asbattha
- **Gujarati:** Piplo
- **Hindi:** Pipal
- **Irula:** Arasa
- **Kannada:** Arali
- **Malayalam:** Aal, Ashvatham, Bodhivriksham
- **Manipuri:** Sana Khongnang
- **Oriya:** Aswattha

Season: Flowering and Fruiting: May to July

Parts used: All parts

Characteristics:

1. Size and Growth Habit: *F. religiosa* is a tall and majestic tree that can reach heights up to 30 meters (98 feet) or more. It has a spreading canopy with a wide crown and a straight, cylindrical trunk (Sajwan *et al.*, 1977; Warrier, 1996; Panchawat, 2012; Bhalerao *et al.*, 2014).

2. Leaves: The leaves of *F. religiosa* are distinct and have a heart-shaped or broadly ovate shape. They are alternate, simple, and have a long, pointed tip. The leaves have prominent veins that radiate from the base and a smooth margin. They are typically dark green and have a glossy texture (Sajwan *et al.*, 1977; Warrier, 1996; Panchawat, 2012; Bhalerao *et al.*, 2014) (Fig. **28.2**).

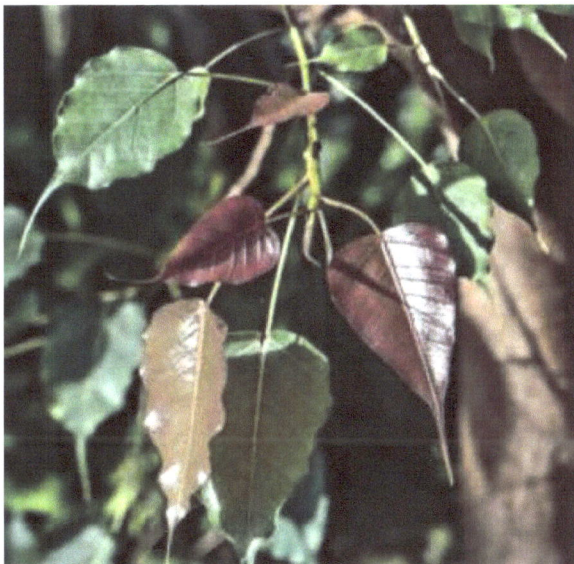

Fig. (28.2). Leaves of *F. religiosa* (PC: Apurva Shankar Chonde).

3. Bark: The bark of *F. religiosa* is smooth and grayish in color when the tree is young. As the tree ages, the bark becomes rougher, developing a mottled appearance with patches of light gray, brown, and off-white colors (Sajwan *et al.*, 1977; Warrier, 1996; Panchawat, 2012; Bhalerao *et al.*, 2014).

4. Aerial Roots: One of the most prominent features of *F. religiosa* is its aerial roots. These roots emerge from the branches and hang down towards the ground, giving the tree a unique and distinctive appearance. The aerial roots can sometimes reach the ground and develop into secondary trunks, creating a characteristic prop root structure (Sajwan *et al.*, 1977; Warrier, 1996; Panchawat, 2012; Bhalerao *et al.*, 2014).

5. Fruits: The fruits of *F. religiosa* are small, rounded structures known as figs. These figs are actually inverted flowers, with tiny flowers enclosed within the fleshy, pear-shaped structure. When ripe, the figs turn from green to yellow or purple. They are an important food source for birds and other wildlife (Sajwan *et al.*, 1977; Warrier, 1996; Panchawat, 2012; Bhalerao *et al.*, 2014).

Distribution:

Indian subcontinent – Bangladesh, Nepal, Pakistan, and India including the Assam region, Eastern Himalaya, and the Nicobar Islands; Indochina – the Andaman Islands, Myanmar and Peninsular Malaysia; Iran, Florida and Venezuela.

Propagation: Seed, Air layering, Tip cuttings

Chemical constituents:

The bark of this tree contains lanosterol, β –sitosteryl-D glucoside, bergaptol, bergapten, steroids, flavonoids, alkaloids, and phenol content. Campestrol, stigmasterol, isofucosterol, tannins, arginine, serine, aspartic acid, glycine, threonine, alanine, proline, tryptophan, tyrosine, methionine, valine, isoleucine (Sandeep *et al.*, 2018). asparagine, tyrosine, undecane, tridecane, tetradecane, ocimene, limonene, dendrolasine, and flavonoids (Rutuja *et al.*, 2015).

Recipe:

Ingredients: Leaves, onion, ground garlic, green chilies, ginger paste, mustard, cumin seeds, and dry masala, oil, salt, kokum, *etc.*

Method: First, clean and wash the green leaves. Drain out excess water and chop them, then keep them aside. Heat oil in a pan and add mustard and cumin seeds. Once they splutter, add the chopped onions and sauté them until they become translucent. Next, add the chopped leaves and mix them well. Add dry masala, salt, and kokum. Cover and cook for a few minutes. It is now ready to be served with rotis or bhakri, along with salad.

Uses:

1. Religious and Cultural Significance: *F. religiosa* is considered sacred and revered in several religions, particularly Buddhism. It is believed to be the tree under which Gautama Buddha attained enlightenment. As a result, it is often planted near temples, monasteries, and other religious sites. People pay homage to the tree and often meditate or seek spiritual solace under its shade (Bhalerao *et al.*, 2014).

2. Shade and Ornamental Tree: Due to its large size and spreading canopy, *F. religiosa* is often planted as a shade tree in parks, gardens, and along streets. Its wide crown provides excellent shade, making it a preferred tree in hot climates. Additionally, the tree's unique appearance and cultural significance make it a popular choice for ornamental purposes in landscapes and gardens.

3. Environmental Benefits: *F. religiosa* has significant ecological value. The dense foliage of the tree provides habitat and food for various bird species, insects, and other small animals. Its large canopy helps regulate temperature and humidity, reducing the impact of urban heat islands (Singh *et al.*, 2014).

4. Medicinal: In traditional medicine, various parts of *F. religiosa*, including the bark, leaves, and roots, are used for their medicinal properties. They showed anti-inflammatory, antimicrobial, and antioxidant effects. The tree's extracts and preparations are used to treat various ailments, including gastrointestinal disorders, respiratory issues, skin diseases, and diabetes. Leaves are used in conditions like vomiting, gonorrhea, *etc.* and can also be consumed in the form of juice in conditions like asthma, cough, and diarrhea (Panchawat, 2012; Singh *et al.*, 2014).

5. Timber and Woodcraft: The wood of *F. religiosa* is moderately hard and durable. It is used in the construction of furniture, doors, windows, and musical instruments. The tree's timber is also employed in carving, turning, and other forms of woodcraft (Singh *et al.*, 2014).

CHAPTER 6

Wild Vegetables of the Family Musaceae

INTRODUCTION

The Musaceae family, also known as the banana family, has unique morphological characteristics such as large herbaceous growth, pseudo stem, large elongated leaves with prominent midrib and entire margin, berry fruits, and spadix inflorescence covered with spathe. It consists of 6 genera and 130 species (Qamar and Shaikh 2011). The members are rich in carbohydrates in the form of simple sugars (fructose, sucrose, glucose, *etc.*) and starch, vitamin B, Minerals-potassium, magnesium, manganese, phosphorus, and pectin fibers. Some members contain antinutritional compounds such as tannins, lactins, phytic acid, *etc.* (Qamar and Shaikh 2011).

Ensete superbum (Roxb.) Cheesman

Botanical name: *Ensete superbum*

Family: Musaceae

Local name: Raankel, Chaveni

Vernacular name:

- **Assamese:** Lobong keng tong
- **Gujarati:** Jangli kela
- **Hindi:** Jangli kela
- **Kannada:** Bettabale, Kaadubale, Kallubale
- **Konkani:** Jangli keli, Raan kyaanle
- **Malayalam:** Kalluvazha, Kattuvazha, Malavazha
- **Marathi:** Chaveni, Kavadara, Raankel
- **Rajasthani:** Jangli kelo
- **Sanskrit:** Bahubija
- **Tamil:** Kal-valai, Kattu-valai, Malai-valai
- **Telugu:** Adavi arati
- **Tulu:** Kallubare

Season: June to August

Parts used: whole plant

Characteristics (Vasundharan *et al.*, 2015; Sethiya *et al.*, 2019):

1. Plant Size and Habit: *E. superbum* is a large perennial herbaceous plant with a pseudostem that can reach a height of up to 10 meters (33 feet) or more. The pseudostem is formed by tightly overlapping leaf sheaths and lacks a true woody stem.

2. Leaves: The leaves of *E. superbum* are large and elongated, typically measuring between 1.5 to 4 meters (5 to 13 feet) in length. They are arranged spirally around the stem and have a prominent midrib. The leaves are leathery in texture and may have a bluish-green color.

3. Inflorescence: *E. superbum* produces a large and showy inflorescence borne on a long, erect stalk. The inflorescence consists of numerous individual flowers that are arranged in a dense, cone-shaped cluster.

4. Flowers: The flowers of *E. superbum* are small and inconspicuous. They are usually yellow or pale green in color. The individual flowers are unisexual, meaning they are either male or female, and are grouped together within the inflorescence.

5. Fruits: After successful pollination, *E. superbum* produces clusters of green to reddish-brown fruits. The fruits are elongated and can measure up to 12 centimeters (4.7 inches) in length. They contain numerous small seeds (Fig. **29.1**).

Distribution:

It is native to the tropical regions of Southeast Asia, including countries like India, Myanmar (Burma), Thailand, and Malaysia. It is commonly found in moist forested areas, including valleys, ravines, and rocky slopes.

Propagation: seeds, corms

Chemical constituents:

Alkaloids, saponins, tannins, phenols, flavonoids, anthraquinone glycosides, steroids and amino acids. (Mishal *et al.*, 2020).

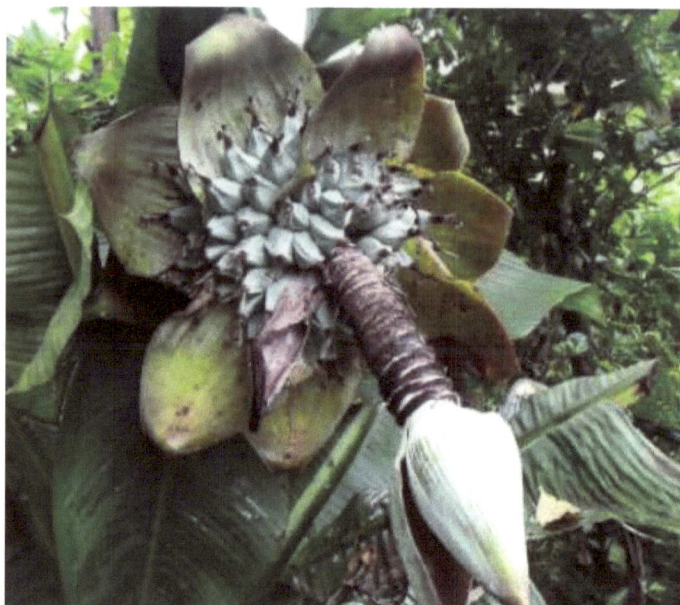

Fig. (29.1). Fruits of *E. superbum* (PC: Apurva Shankar Chonde).

Recipe:

Ingredients: flower, green peas, onion, cumin seeds, hing, jaggery, masala, kokum, salt, *etc.*

Method: Clean the banana blossom or plantain flower and chop the florets finely. Soak them in diluted buttermilk or plain water. Add turmeric and salt to the buttermilk and let it soak overnight in the refrigerator, or for at least 30 minutes. Prepare the Malvani masala as per the recipe and keep it ready. Heat a saucepan with oil, add cumin seeds and mustard, and let them crackle for 10 seconds. Add asafoetida, chopped garlic, and onions, and fry until the onions turn translucent. Once done, add the chopped florets (without the soaking liquid) and the Malvani masala, along with salt to taste. Toss the ingredients together, add jaggery and kokum, and sauté until the florets soften and cook. You can sprinkle some water, cover the pan with a lid, and cook further. Make sure to keep sautéing to prevent sticking at the bottom of the pan. Serve the Maharashtrian Style Kelfulachi Bhaji alongside the Khandeshi Dal Recipe or Khichdi for a wholesome and healthy lunch.

Uses:

1. Ornamental Plant: *E. superbum* is often cultivated as an ornamental plant in gardens and landscapes due to its large, tropical foliage and dramatic appearance.

The impressive size and unique shape of its leaves make it an attractive addition to tropical-themed gardens (Saroj *et al.*, 2012).

2. Medicinal Purposes: In traditional medicine systems, various parts of *E. superbum* are believed to possess medicinal properties. Different preparations of the plant have been used to treat conditions such as diarrhea, dysentery, wounds, and fever. Seeds are especially used in the treatment of diabetes, kidney stone and leucorrhoea (Sreedharan *et al.*, 2004; Yesodharan *et al.*, 2007; Udayan *et al.*, 2008; Ahirrao *et al.*, 2010; Mohan *et al.*, 2010).

3. Culinary: In some cultures, the young shoots of *E. superbum* are consumed as a vegetable. The tender shoots are cooked and used in various dishes, similar to other edible banana plant species. However, it is important to note that consuming any plant parts should be done with caution, and proper identification and preparation methods are essential (Arinathan *et al.*, 2007; Jagtap *et al.*, 2008; Sethiya *et al.*, 2019).

4. Fiber and Craft: The fibers obtained from the pseudostem of *E. superbum* have been used for making ropes, baskets, and other crafts in certain regions. The durable and strong nature of the fibers makes them suitable for various practical applications (Jagtap *et al.*, 2008).

Musa paradisiaca L.

Botanical name: *Musa paradisiaca*

Family: Musaceae

Local name: Keli, Banana

Vernacular name:

- **Hindi:** Kela
- **Kannada:** Baazhe
- **Malayalam:** Vaazha
- **Manipuri:** Laphoo tharo
- **Tamil:** Vaazha
- **Nepali:** Kera
- **Telugu:** Arati
- **Mizo:** Balhlakual, Vaibalhla
- **Angami:** Tekhusi
- **Sanskrit:** Kadalee, Vaaranabusaa, rambhaa, Mochaa, Amshumatphalaa

Season: July to August

Parts used: Fruits

Characteristics (Galani 2019):

1. Plant Type: *M. paradisiaca* is an herbaceous plant that belongs to the family Musaceae. It has a large, upright pseudostem (stem-like structure) that grows from an underground rhizome. The plant can reach heights of 3 to 9 meters, depending on the variety (Fig. **29.2**).

Fig. (29.2). Fruits of *M. paradisiaca* (PC: Tukaram Kanoja).

2. Leaves: The leaves of *M. paradisiaca* are large and elongated, with a characteristic oblong shape. They arise from the top of the pseudostem and have a smooth, waxy surface. The leaves are arranged spirally and have prominent veins running through them.

3. Inflorescence: *M. paradisiaca* produces a terminal inflorescence known as a banana bunch or cluster. The inflorescence is pendulous and consists of several tightly packed flowers called bracts. The bracts are thick and fleshy, usually green or purple in color, and gradually change into bananas as they mature.

4. Fruit: The fruit of *M. paradisiaca*, commonly referred to as a banana, is the main distinguishing feature. Bananas are elongated, fleshy berries with a curved

shape. They vary in size, color, and texture depending on the variety. The skin of the banana is usually yellow when ripe but can also be green, red, or even striped in some varieties (Fig. **29.2**).

5. Rhizome: *M. paradisiaca* has a rhizomatous root system, where the underground rhizome gives rise to new shoots and stems. The rhizome is a thick, fleshy structure that stores nutrients and energy for the plant.

Distribution:

India, Ethiopia, Myanmar, Thailand and Vietnam.

Propagation: Corms

Chemical constituents:

Carbohydrates (Sugars)- glucose, fructose, and sucrose, dietary fiber, Vitamins-vitamin C, vitamin B6, and folate (vitamin B9), minerals, potassium, magnesium, and manganese, phenolic compounds- catechins, flavonoids, and dopamine, tryptophan, pectin.

Recipe:

Ingredients:flour, green peas, onion, cumin seeds, hing, jaggery, masala, kokum, and salt.

Method: Clean the banana blossom/plantain flower. Finely chop the florets and soak them in diluted buttermilk or plain water. Add turmeric and salt to the buttermilk and allow it to soak overnight (refrigerated) or for at least 30 minutes. Prepare the Malvani masala as per the recipe and keep it ready. Heat a saucepan with oil, add cumin seeds and mustard, and let them crackle for 10 seconds. Add hing, chopped garlic, and onions, and fry until the onions turn translucent. Once done, add the chopped florets (without the water). Add the Malvani masala and salt to taste. Toss it well, add jaggery and kokum, and sauté until the florets soften and cook. You can sprinkle some water, cover the pan with a lid, and cook further. Remember to keep stirring to prevent it from sticking to the bottom of the pan. Serve the Maharashtrian Style Kelfulachi Bhaji with Khandeshi Dal Recipe or Khichdi for a delicious and healthy lunch.

Uses:

1. Food: The most significant use of *M. paradisiaca* is as a food source. Bananas are consumed worldwide and are enjoyed both raw and cooked. They can be eaten as a standalone fruit, added to smoothies, used in desserts, or included in various

culinary preparations. Bananas are known for their natural sweetness, creamy texture, and nutritional value (Gul *et al.* 2016; Patiño-Rodríguez *et al.* 2019).

2. Nutritional Benefits: *M. paradisiaca* is a nutritious fruit. It is a good source of vitamins, including vitamin C, vitamin B6, and folate, as well as minerals like potassium and magnesium. Bananas also contain dietary fiber, which aids in digestion and promotes satiety (Kemisetti *et al.*, 2022).

3. Energy Boost: Bananas are a popular choice for athletes and individuals seeking a quick energy boost. They provide a natural source of carbohydrates, including sugars like glucose and fructose, which are easily digested and converted into energy (Kemisetti *et al.*, 2022).

4. Cooking Ingredient: Bananas can be used as an ingredient in various recipes. They can be used to add natural sweetness, moisture, and flavor to baked goods like bread, muffins, and cakes. Bananas can also be used in savory dishes, such as curries and stews, to impart a unique flavor and texture (Kemisetti *et al.*, 2022).

5. Banana Leaves: The large, flexible leaves of *M. paradisiaca*, known as banana leaves, have culinary and practical uses. In some cuisines, banana leaves are used as natural food wrappers for steaming or grilling dishes. They can also be used as plates or serving surfaces for traditional meals. Additionally, banana leaves are used for various purposes such as food storage, wrapping, and as eco-friendly packaging material (Kemisetti *et al.*, 2022).

CHAPTER 7

Wild Vegetables of the Family Nelumbonaceae

INTRODUCTION

This family is also known as the Lotus family with unique features such as aquatic habitat, large peltate and floating leaves, and flowers- large, solitary, conspicuous with numerous petals with colors ranging from white, yellow, pink or red. It consists of one genus and two species (Xue *et al.*, 2012). The members are a rich source of carbohydrates, proteins, and minerals including potassium, phosphorus, manganese, magnesium, iron, vitamins, and fibers. However, they also possess antinutritional contents such as nelumbone alkaloids and tannins (Sujitha *et al.*, 2013).

Nelumbo nucifera Gaertn.

Botanical name: *Nelumbo nucifera*

Family: Nelumbonaceae

Local name: lotus

Vernacular name:

- **Tamil:** Ambal, Thamarai, Padma, Pankaja
- **German:** Indischelotosblume
- **Gujrat:** Suriyakamal
- **Malayalam:** Tamara
- **Bengal:** Padma
- **French:** Nelumbo
- **Sanskrit:** Ambuja
- **Hindi:** Kanwal, Kamal
- **Persian:** Nilufer

Season: April to September

Parts used: Flower, Stem

Characteristics:

1. Aquatic Plant: *N. nucifera* is an aquatic perennial plant that grows in freshwater environments, such as ponds, lakes, and slow-moving rivers. It has adapted to thrive in wetland habitats (Mukherjee *et al.*, 1996) (Fig. **30.1**).

(a)   (b)

Fig. (30.1). *N. nucifera* **a) whole plant b) rhizome** (PC: Apurva Shankar Chonde).

2. Rhizomatous Growth: The plant has a thick, fleshy rhizome (underground stem) that anchors it in the soil or mud at the bottom of water bodies. From the rhizome, long petioles (stalks) and leaves emerge above the water surface (Mukherjee *et al.*, 1996) (Fig. **30.1**).

3. Leaves: The leaves of *N. nucifera* are large, round, and circular in shape. They are borne on long, sturdy petioles that rise above the water. The leaves can reach a diameter of up to 60 centimeters. The upper surface of the leaves is smooth, while the lower surface is covered with tiny hairs (Mukherjee *et al.*, 1996).

4. Flowers: *N. nucifera* produces beautiful, showy flowers that are considered sacred and culturally significant in many regions. The flowers are large, with a diameter of about 20 to 25 centimeters. They have numerous petals arranged in multiple layers, giving them a distinctive appearance. The petals can be white, pink, or pale yellow, depending on the variety. The flowers emit a pleasant fragrance (Mukherjee *et al.*, 1996) (Fig. **30.1**).

5. Fruits: The fruits are rounded or oval in shape and have a unique appearance. They consist of multiple chambers, each containing a single seed embedded in a

hard, woody shell. The fruits are typically green when young and turn brown or black as they mature (Mukherjee *et al.*, 1996).

6. Cultural and Symbolic Significance: *N. nucifera* holds great cultural and religious significance in various societies. It is often associated with purity, enlightenment, and spiritual awakening. The plant is widely featured in art, literature, and religious practices.

Distribution:

Asia- India, China, Japan, Sri Lanka, Thailand, Vietnam, Cambodia, Myanmar, and Bangladesh, North America, Australia, Africa, Europe.

Propagation: Seed, dividing rhizomes

Chemical constituents:

Anonaine, ß-sitosterol, Dauricine, Demethylcoclaurine, Dehydroanonaine, Ginnol, Gluconic acid, Hyperin, Kaempferol-3-O-ß-glucuronide, Linalool, Lotusine, Luteolin, Luteolin glucoside, Malic acid, N-noramepavine, Nornuciferine, Nuciferine, Pronuciferin, Quercetin, Roemerrine, Rutin, Tartaric acid (Paudel *et al.*, 2015).

Recipe:

Ingredients: Lotus stem, onions, curd, ginger-garlic paste, coriander powder, turmeric powder, red chili powder, salt and cumin powder, water, *etc.*

Method: Put the lotus stem slices in a pressure cooker and add 1 cup of water. Cook them until they become soft. Cook them for one whistle on high heat and two whistles on low heat. Then, add onions, ginger-garlic paste, curd, coriander powder, turmeric powder, red chili powder, salt, and cumin powder. Add half a cup of water and fry the masala until the oil separates on the side of the pan. Next, add the cooked lotus stems along with the water. Cover and cook for 10 minutes. Remove the lid and increase the heat to high. Cook until all the water has evaporated and the vegetable becomes dry and nicely cooked.

Uses:

1. Cultural and Spiritual Significance: *N. nucifera* holds great cultural and religious significance in many societies, particularly in Asia. It is considered a sacred symbol of purity, enlightenment, and spiritual awakening. The lotus is often depicted in religious artwork, architecture, and ceremonies (Shen-Miller, 2002).

2. Ornamental Plant: *N. nucifera* is widely cultivated and appreciated for its beautiful flowers and leaves. It is commonly grown in water gardens, ponds, and containers as an ornamental plant. The showy and fragrant flowers of the lotus add a touch of elegance and beauty to landscapes and gardens (Wang *et al.*, 2005).

3. Medicinal Purposes: Various parts of *N. nucifera*, including the flowers, leaves, seeds, and rhizomes, have been used in traditional medicine practices. It possesses therapeutic properties and has been used to address various health concerns. The lotus is often associated with benefits such as improving digestion, promoting relaxation, reducing inflammation, and enhancing skin health (Wang *et al.*, 2005; Sridhar *et al.*, 2007; Sujitha *et al.*, 2013).

4. Culinary: In some cultures, the seeds, rhizomes, and young leaves of *N. nucifera* are used in culinary preparations. The lotus seeds, also known as "makhana," are consumed as a snack or used in sweet and savory dishes. The rhizomes, known as lotus roots, are used in soups, stir-fries, and other dishes for their crunchy texture and mild flavor (Sridhar *et al.*, 2007).

5. Herbal Tea: Lotus flower tea is popular in many cultures. The dried petals or stamens of *N. nucifera* are steeped in hot water to make a fragrant and soothing herbal tea. Lotus tea is often enjoyed for its calming properties and delicate floral aroma (Sridhar *et al.*, 2007).

6. Cosmetics and Skincare: Extracts and oils derived from *N. nucifera* are used in cosmetic and skincare products. The lotus is known for its moisturizing and nourishing properties, and it is used in creams, lotions, soaps, and hair care products to promote healthy skin and hair (Sridhar *et al.*, 2007).

CHAPTER 8

Wild Vegetables of the Family Nyctaginaceae

INTRODUCTION

The Nyctaginaceae family, also known as the four-o'clock family consists of 33 genera and 290 species (Xu and Deng 2017). Some members contain alkaloids, phenolic compounds, flavonoids, glycosides, tannins, saponins, *etc.* (Pooja *et al.*, 2017).

Boerhavia diffusa L.

Botanical name: *Boerhavia diffusa* L.

Family: Nyctaginaceae

Local name: Gadapushpa, Punarnava

Vernacular name:

- **Common name:** Red hogweed, Tar Vine, Red Spiderling, Wineflower
- **Hindi:** Punarnava, Satha
- **Kannada:** Adakaputta, Adakaputtana gida, Komme, Gonajaali, Sanaadikaa Balavadike, Belavadaka, Shavaata, Shivaata, Shivaatike, Nadumurukana balli
- **Nepali:** Punarnavaa, Punarvaa, Laal Punarnavaa, Saano Paate, Laal Gaj Purnee, Aule Saag
- **Sanskrit:** Punarnavaa

Season: Flowering: April – June; Fruit ripening: July - August

Parts used: Whole plant

Characteristics:

1. Growth Habit: It is a perennial herbaceous plant that typically grows upright or prostrate. It can reach a height of up to 1 meter (Thakur *et al.*, 1989).

2. Leaves: The leaves are simple, alternate, and usually have a long petiole. The leaf shape can vary, but they are generally ovate to lanceolate with pointed tips. The leaf margins are often smooth or slightly toothed (Thakur *et al.*, 1989).

3. Flowers: *B. diffusa* produces small, pink or white flowers that are arranged in clusters or spikes. The flowers have a five-lobed tubular structure and are often fragrant (Thakur *et al.*, 1989) (Fig. **31.1**).

 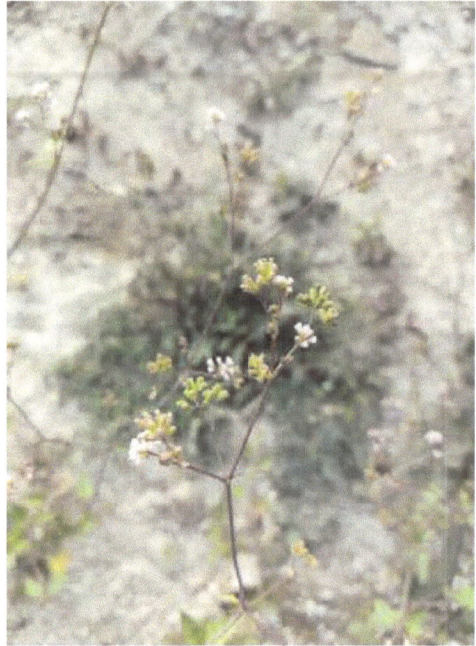

(a)   (b)

Fig. (31.1). *B. diffusa* **a) Whole plant b) Inflorescence** (PC: Ganesh Pawar).

4. Root System: The roots are fibrous and can penetrate deep into the soil. They may have a reddish-brown color (Thakur *et al.*, 1989).

Distribution:

Africa- Nigeria, Ghana, Sudan, Ethiopia, Tanzania, and South Africa, Asia- India, Pakistan, Bangladesh, Sri Lanka, Myanmar, Thailand, Malaysia, Indonesia, and the Philippines, Americas- Central America, South America, and the Caribbean, Other countries- Mexico, Brazil, Colombia, Ecuador, Peru, Venezuela, and Jamaica.

Propagation: Seed

Chemical constituents:

Tannins, flavonoids, alkaloids (punarnavine), glycosides, steroids, terpenoids, phenolic compounds, rotenoids (Shisode *et al.*, 2011); phenolic compounds, saponins, glycosides (Pooja *et al.*, 2017).

Recipe:

Ingredients: Punarnava bunches, toor dal, tomatoes, onion, green chilies, curry leaves, ginger garlic paste, turmeric powder, red chili powder, water, salt, coriander leaves, mustard seeds, cumin seeds.

Method: In a pressure cooker, add cleaned and washed Punarnava (Galijeru aaku), onions, toor dal, green chilies, curry leaves, ginger-garlic paste, turmeric powder, tomatoes, red chilies, and oil. Mix well and pressure cook the dal for about 2 whistles. Once the pressure is completely released, mash the dal and leaves, and mix well. In another pan, add oil along with mustard seeds, cumin seeds, dry red chilies, crushed garlic, and turmeric powder. Mix well and cook until the tempering releases its aroma. Then, add this tempering to the cooked dal and mix well. Cook for about 2 minutes. Finally, add coriander leaves and mix well. Serve hot with plain rice or even with roti.

Uses:

1. Nutrient Content: *B. diffusa* is reported to contain various nutrients such as proteins, carbohydrates, vitamins (including vitamin C), and minerals (including calcium, iron, and potassium) (Nayak *et al.*, 2016).

2. Diuretic Activity: *B. diffusa* is traditionally known for its diuretic properties. It may increase urine output and promote the elimination of excess fluids from the body (Nayak *et al.*, 2016).

3. Anti-inflammatory Activity: *B. diffusa* possesses anti-inflammatory properties. It may help reduce inflammation and provide relief in conditions such as arthritis and inflammatory disorders (Mudgal *et al.*, 1974).

4. Antioxidant Activity: It has been found to exhibit antioxidant properties, which can help protect against oxidative stress and damage caused by free radicals in the body (Nayak *et al.*, 2016).

5. Hepatoprotective Activity: Plants may possess hepatoprotective properties, potentially helping to protect the liver from damage and support its overall health (Muzila, 2006).

6. Anti-diabetic Activity: *B. diffusa* may have anti-diabetic effects by influencing blood glucose levels and insulin secretion. It may potentially be used as a complementary therapy for managing diabetes (Nayak *et al.*, 2016).

7. It is useful in strangury, leucorrhoea, ophthalmia, lumbago, myalgia, cardiac disorders, jaundice, anaemia, dyspepsia, constipation, cough, bronchitis, and general debility (Prajapati *et al.*, 2004). Roots are useful for gonorrhoea, dropsy, bronchial asthma, night blindness, rheumatism, and several diseases of urine, liver, kidney, fever, and heart (Qureshi *et al.*, 2001; Singh *et al.*, 2002; Parveen *et al.*, 2007; Jaiswal, 2010).

CHAPTER 9

Wild Vegetables of the Family Oleaceae

INTRODUCTION

The Oleaceae family is distinctive, typically consisting of trees or shrubs (rarely lianas), usually featuring peltate secretory trichomes and opposite leaves. Their inflorescences are typically cymes or solitary flowers (Green 2004). It consists of 25 genera and 688 species (Huang *et al.*, 2019). Some members are a valuable source of healthy fats like oleic acid. However, it may also contain a bitter compound *i.e.* oleuropein.

Schrebera swietenioides Roxb.

Botanical name: *Schrebera swietenioides*

Family: Oleaceae

Local name: Mokha

Vernacular name:

- **Common name:** Weaver's Beam Tree
- **Hindi:** Banpalas, Mokhdi, Mokha
- **Kannada:** Bula, Gante, Mogalingamara
- **Malayalam:** Maggamaram, Malamplasu, Muskkakavrksam
- **Marathi:** Mokha, Mokadi, Nakti
- **Oriya:** Mokka
- **Sanskrit:** Ghantapatali, Golidha, Kastapatola
- **Tamil:** Kattupparutticceti, Mogalingam, Makalinkam
- **Telugu:** Bullakaya, Magalinga, Tondamukkudi

Season: Flowering and fruiting: May-June

Parts used: Roots, Bark, Leaves, Fruits, Alkali

Ganesh Chandrakant Nikalje, Apurva Chonde, Sudhakar Srivastava & Penna Suprasanna

Characteristics (Green 2004):

1. Habitat: This species is typically found in evergreen and deciduous forests, as well as in moist or dry areas. It can be found at elevations ranging from sea level to around 1,500 meters (4,900 feet).

2. Tree: *S. swietenioides* is a medium to large-sized tree that can reach heights of up to 25 meters (82 feet). It has a straight, cylindrical trunk with a rough, fissured bark.

3. Leaves: The leaves are simple, opposite, and usually clustered at the end of branches. They are elliptical or lanceolate in shape, with a glossy green color. The leaf margins are serrated or toothed.

4. Flowers: The tree produces small, fragrant flowers that are borne in axillary or terminal panicles. The flowers have a tubular shape and are usually white or cream-colored. They have four or five petals and are bisexual.

5. Fruits: After successful pollination, *S. swietenioides* produces fleshy, ovoid or ellipsoid fruits. The fruits are initially green and turn yellow or orange when ripe. Each fruit contains a single seed (Fig. **32.1**).

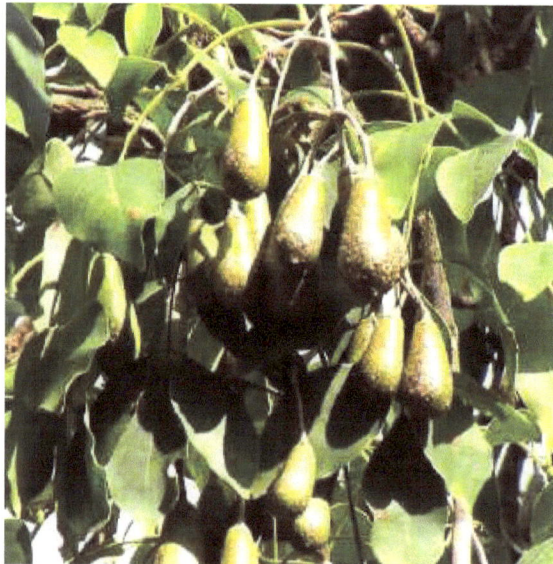

Fig. (32.1). Fruits of *S. swietenioides* (PC: Madhukar).

6. Seeds: The seeds of *S. swietenioides* are round or oval and have a hard, woody coat. They are dark brown or black in color.

Distribution:

India- Western Ghats, Eastern Ghats, Kerala, Karnataka, Tamil Nadu, Maharashtra, and Andhra Pradesh; Sri Lanka, Myanmar, Thailand.

Propagation: Seed

Chemical constituents:

Alkaloids- schreberine, schreberamine, and schreberamine N-oxide, triterpenoids-betulinic acid, oleanolic acid, and ursolic acid, flavonoids- apigenin, luteolin, and quercetin, sterols- β-sitosterol, stigmasterol, essential oils.

Recipe:

Ingredients: Younger leaves, finely chopped onion, 5-6 garlic cloves, salt, green chilies, curry leaves, Asafoetida, mustard seeds, turmeric, gram dal, *etc.*

Method: Wash the leaves and chop them finely. Heat oil in a pan and fry asafoetida, mustard seeds, curry leaves, crushed garlic cloves, and finely chopped onion. When the onion turns slightly yellow, add the chopped vegetables, chili, turmeric powder, soaked dal, and salt to taste. Lightly beat water, cover, and cook over low heat.

Uses:

1. Culinary: The leaves can be used as vegetables (Manda *et al.*, 2009).

2. Medicinal: Various parts of *S. swietenioides*, including the leaves, bark, and roots, have been used in traditional medicine. It possesses medicinal properties such as antipyretic (fever-reducing), anti-inflammatory, analgesic (pain-relieving), and antihelmintic (against worms) effects. It has been used to treat conditions like fever, rheumatism, arthritis, skin diseases, gastrointestinal disorders, and respiratory ailments (Manda *et al.*, 2009; Rasal *et al.*, 2009).

3. Timber: The wood of *S. swietenioides* is used for timber purposes. It is valued for its durability and resistance to termites and is used in construction, carpentry, and furniture making (Rasal *et al.*, 2009).

4. Ornamental Purposes: The attractive foliage and fragrant flowers of *S. swietenioides* make it suitable for ornamental purposes. It is sometimes grown as an ornamental tree in gardens and parks for its aesthetic value (Rasal *et al.*, 2009).

CHAPTER 10

Wild Vegetables of the Family Onagraceae

INTRODUCTION

The Onagraceae, commonly referred to as the evening-primrose or willowherb family, comprises around 17 genera and 650 species of flowering plants, including herbs, shrubs, and trees (Shawky *et al.*, 2021). The edible members are a good source of Vitamin A and C, Gamma-Linolenic Acid, minerals, fibers, *etc.* However, some members contain oxalates, and cyanogenic glycosides, which are anti-nutritional compounds (Shawky *et al.*, 2021).

Epilobium angustifolium L.

Botanical name: *Epilobium angustifolium*

Family: Onagraceae

Local name: Fireweed

Common name: Rosebay Willow-Herb, Fireweed

Season: June to September

Parts used: Leaves, flowers, roots

Characteristics:

1. Plant Height and Growth Habit: *E. angustifolium* is a tall perennial herbaceous plant that can reach heights of up to 1.5 to 2 meters (5 to 6.5 feet). It has an erect and branching growth habit (Myerscough, 1980; Broderick, 1990; Buchwald *et al.*, 2006; Wiese *et al.*, 2012).

2. Leaves: The leaves of *E. angustifolium* are narrow and lance-shaped, giving rise to its specific epithet "angustifolium," which means "narrow-leaved." The leaves are alternate, simple, and usually have smooth margins. They are typically between 5 to 10 centimeters (2 to 4 inches) long (Myerscough, 1980; Broderick, 1990; Buchwald *et al.*, 2006; Wiese *et al.*, 2012).

Ganesh Chandrakant Nikalje, Apurva Chonde, Sudhakar Srivastava & Penna Suprasanna

3. Stem: The stems of *E. angustifolium* are erect, slender, and usually reddish or purplish in color. The stems are typically hairy and may have fine hairs or glandular pubescence.

4. Flowers: *E. angustifolium* produces showy, terminal spikes of flowers. The flowers are usually pink to magenta in color, but can also be white or purplish. Each flower has four petals and four sepals. The petals are notched at the tips, giving them a fringed appearance (Myerscough, 1980; Broderick, 1990; Buchwald *et al.*, 2006; Wiese *et al.*, 2012) (Fig. **33.1**).

Fig. (33.1). **Inflorescence of *E. angustifolium*** (PC: Gaytri Chonde).

5. Inflorescence: The flowers of *E. angustifolium* are arranged in dense, elongated clusters or spikes at the top of the stem. The inflorescence can be quite tall and can contain numerous individual flowers (Myerscough, 1980; Broderick, 1990; Buchwald *et al.*, 2006; Wiese *et al.*, 2012).

6. Fruits: After flowering, *E. angustifolium* produces slender seed capsules that contain numerous small seeds. The capsules split open to release the seeds, which are dispersed by wind (Myerscough, 1980; Broderick, 1990; Buchwald *et al.*, 2006; Wiese *et al.*, 2012).

Distribution:

It is a widespread plant found in various habitats, including open meadows, roadsides, disturbed areas, and burned areas. It is native to northern regions of the Northern Hemisphere, including North America, Europe, and Asia.

Propagation: Seed

Chemical constituents:

Flavonoids- quercetin, kaempferol, and their glycosides, tannins, phenolic acids- gallic acid, ellagic acid, and caffeic acid, phytosterols- beta-sitosterol, essential oils containing linalool, geraniol, and alpha-pinene.

Recipe:

Ingredients: Younger leaves, finely chopped onion, 5-6 garlic cloves, salt, green chilies, curry leaves, Asafoetida, mustard seeds, turmeric, gram dal, *etc.*

Method: Peel the stalks of the leaves and wash the young leaves. Chop them finely. Heat oil in a pan and fry asafoetida, mustard seeds, curry leaves, crushed garlic cloves, and finely chopped onion. When the onion turns slightly yellow, add the chopped vegetables, chili, turmeric powder, soaked dal, and salt to taste. Lightly beat some water, cover the pan, and cook over low flame.

Uses:

1. Medicinal Purposes: *E. angustifolium* has a long history of traditional use in herbal medicine. It is believed to have anti-inflammatory, anti-allergic, anti-atherogenic, anti-microbial, anti-viral, anti-proliferative, and immunomodulatory properties. It has been used in the treatment of various conditions, including gastrointestinal disorders, urinary tract infections, prostate issues, skin irritations, and respiratory ailments (Vitalone *et al.*, 2001; Feldman, 2005; Okuda, 2005; Holderness *et al.*, 2008; Stagos *et al.*, 2012; Vogl *et al.*, 2013; Gollucke *et al.*, 2013; Korkina *et al.*, 2013; Chirumbolo, 2014; Ratz-Lyko *et al.*, 2015).

2. Herbal Tea: The leaves and flowers of *E. angustifolium* can be used to prepare herbal tea. The tea is often enjoyed for its mild flavor and potential health benefits. It is sometimes used as a calming and soothing beverage (Kalle *et al.*, 2020).

3. Culinary: The young shoots and leaves of *E. angustifolium* are edible and can be used as a food source. They can be added to salads, soups, stir-fries, or used as a green vegetable in various culinary preparations (O'Keefe, 2016).

4. Bee Forage: The flowers of *E. angustifolium* are a rich source of nectar and pollen, making them attractive to bees and other pollinators. It is often cultivated to support honeybee populations and promote pollinator-friendly habitats (Shawky *et al.*, 2021).

5. Ornamental Plant: *E. angustifolium* is sometimes cultivated as an ornamental plant for its attractive flowers and foliage. Its tall, colorful spikes of flowers can add beauty to gardens and landscapes (Shawky *et al.*, 2021).

CHAPTER 11

Wild Vegetables of the Family Orchidaceae

INTRODUCTION

The Orchidaceae family, commonly known as the orchid family, is celebrated for its remarkable diversity and beauty. It is one of the largest families in angiosperms with about 850 genera and 20000 species (Gantait *et al.*, 2021). Some members contain carbohydrates, proteins, fiber, and essential minerals such as potassium, calcium, zinc, copper, iron, *etc.* In addition, some members contain antinutritional compounds such as phytates, oxalates, and condensed tannins.

Nervilia concolor (Blume) Schltr.

Botanical name: *Nervilia concolor*

Family: Orchidaceae

Local name: ekpani, duduki

Season: May - September

Parts used: leaves, tubers

Characteristics (Lipińska *et al.*, 2022):

1. Leafless Terrestrial Orchid: *N. concolor* is a leafless orchid species that grows on the ground. Unlike most orchids, it lacks prominent leaves and instead relies on underground structures for nourishment (Fig. **34.1**).

2. Underground Tubers: The plant has fleshy, bulbous underground tubers that serve as storage organs. These tubers are typically rounded or oblong in shape and help the plant survive during periods of dormancy or adverse conditions.

3. Flowering Stem: From the underground tubers, *N. concolor* produces a single flowering stem that rises above the ground. The stem is slender and erect, usually reaching a height of around 10 to 20 centimeters.

Ganesh Chandrakant Nikalje, Apurva Chonde, Sudhakar Srivastava & Penna Suprasanna

Fig. (34.1). Aerial parts of *N. concolor* (PC: Dnyanesh Kamkar).

4. Inflorescence: At the top of the flowering stem, *N. concolor* produces a solitary flower or occasionally a small cluster of flowers. The flower is relatively small, typically measuring about 1 to 2 centimeters in diameter.

5. Flower Structure: The flowers of *N. concolor* have a unique structure. They consist of three petals and three sepals that are similar in appearance and often fused together, forming a tubular or hood-like structure. The flower color is typically white, occasionally with a hint of pink or lavender.

6. Lip or Labellum: The most distinctive feature of *N. concolor* flowers is the lip or labellum, which is the modified petal that is different in shape and color from other petals. The labellum is usually broader and more elaborate, often displaying intricate patterns, markings, or fringed edges.

Distribution:

Asia- China, Japan, Taiwan, Thailand, Myanmar, Cambodia, Vietnam, Malaysia, Indonesia, and the Philippines, Australia, the Pacific Islands, and Africa.

Propagation: Seed

Chemical constituents:

Nervisone, 3,5,7-trimethoxyflavone, 3,7-dimethoxy-5-hydroxyflavone, 5,7-dihydroxy-30,40-dimethoxyflavone, 5,7-dimethoxy-40-hydroxyflavone, 5,7-dimethoxyflavone, 5-hydroxy-7-methoxyflavone, rhamnetin, tetramethylscutellarein (4',5,6,7-tetramethoxyflavone) (Ganbold *et al.*, 2019).

Recipe:

Ingredients: Leaves, finely chopped onion, 5-5 garlic cloves, salt, green chilies, curry leaves, Asafoetida, mustard seeds, turmeric, gram dal, *etc.*

Method: Wash the leaves and chop them finely. Heat oil in a pan and fry the asafetida, mustard seeds, curry leaves, crushed garlic cloves, and finely chopped onion. When the onion turns slightly yellow, add the chopped vegetables, chili, turmeric powder, soaked dal, and salt to taste. Lightly beat water, cover the pan, and cook over low heat.

Uses:

1. Ornamental Plant: *N. concolor* is valued for its attractive and delicate flowers. It is cultivated and used as an ornamental plant in gardens, parks, and greenhouses, adding beauty and diversity to floral displays (Kumar and Boopathi, 2018).

2. Traditional Medicine: Some species of *Nervilia*, including *N. concolor*, are used in traditional medicine systems. Different parts of the plant, such as tubers or rhizomes, may be used in herbal remedies for various purposes. It is believed to possess medicinal properties and is used in remedies for conditions like cough, asthma, and rheumatism. (Dohnal, 1977; Raju *et al.*, 2011a, Raju *et al.*, 2011b; Kumar and Boopathi, 2018; Anand and Basavaraju, 2020).

CHAPTER 12

Wild Vegetables of the Family Oxalidaceae

INTRODUCTION

The Oxalidaceae (570 species), commonly known as the wood sorrel family, is a small group of flowering plants with typically trifoliate leaves in clover-like arrangement and thickened leaf bases of leaflets. Many members show "sleep movements" *i.e.*, the leaflets get folded together during the night or in response to stress or touch. Some members are rich in Vitamin C, and potassium. The presence of oxalic acid gives a 'sour' taste to the plant (Christenhusz and Byng 2016).

Oxalis corniculata L.

Botanical name: *Oxalis corniculata*

Family: Oxalidaceae

Local name: Aambushi

Vernacular name:

- **Hindi:** Amrul
- **Kannada:** Pullampuruche, Pullampurachi, Pullampuriche Hulihulise, Neerugoli, Chaangeri
- **Manipuri:** Yensil
- **Tamil:** Paliakiri
- **Bengali:** Amrulshak
- **Malayalam:** Poliyarala
- **Nepali:** Charee Amilo
- **Mizo:** Siakthur
- **Sanskrit:** Changeri

Season: Flowering: June-August. Fruiting: September-October

Parts used: Leaves

Ganesh Chandrakant Nikalje, Apurva Chonde, Sudhakar Srivastava & Penna Suprasanna

Characteristics:

1. Habit: It is a low-growing perennial herb with a prostrate or creeping growth habit. The stems are slender and spread along the ground, often forming dense mats (Hall *et al.*, 1996; Mary *et al.*, 2001; Saha 2017).

2. Leaves: The leaves of *O. corniculata* are trifoliate, meaning they consist of three leaflets. Each leaflet is heart-shaped or clover-like, with smooth margins and a prominent central vein. The leaflets are usually green but can have purple or reddish markings (Hall *et al.*, 1996; Mary *et al.*, 2001; Saha 2017).

3. Flowers: The flowers of *O. corniculata* are small, about 1 centimeter in diameter, and have five bright yellow petals. The petals often have red or purple streaks near the base. The flowers are borne on slender stalks, rising above the leaves (Hall *et al.*, 1996; Mary *et al.*, 2001; Saha 2017) (Fig. **35.1**).

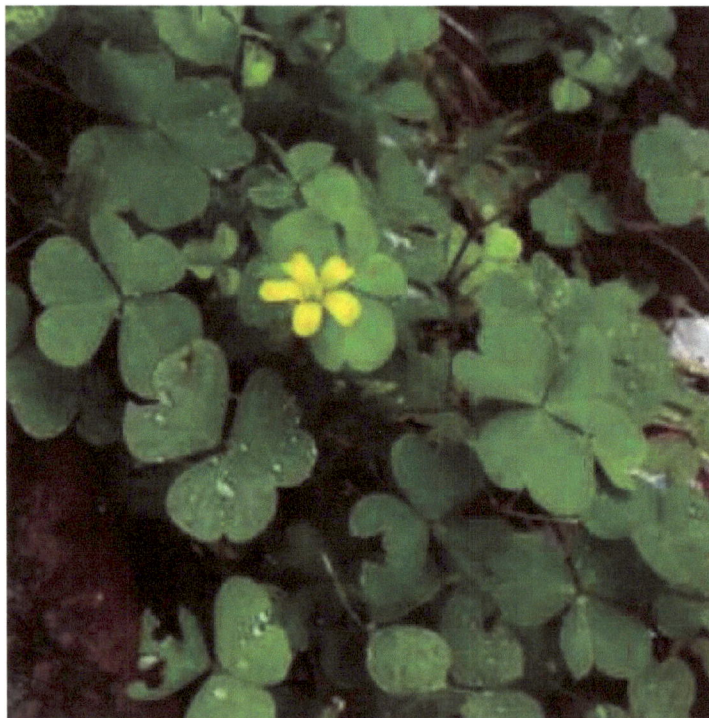

Fig. (35.1). *O. corniculata* **plant with flower** (PC: Apurva Shankar Chonde).

4. Reproduction: *O. corniculata* reproduces both by seeds and by small bulbils that form at the leaf axils. The bulbils can detach from the plant and produce new plants when they come into contact with the ground (Hall *et al.*, 1996; Mary *et al.*, 2001; Saha 2017).

5. Roots: The roots of *O. corniculata* are fibrous and shallow, allowing the plant to spread and colonize various types of soil (Hall *et al.*, 1996; Mary *et al.*, 2001; Saha 2017).

Distribution:

India: Andhra Pradesh, Assam, Bihar, Gujarat, Jammu and Kashmir, Madhya Pradesh, Maharashtra, Manipur, Odisha, Punjab, Rajasthan, Tamil Nadu, Uttar Pradesh; South Europe.

Propagation: Seed

Chemical constituents:

Oxalic acid, flavonoids, anthocyanins, phenolic compounds, *etc.*

Recipe:

Ingredients: Aambushi leaves, Turdal (or green gram, lentils), Danekoot, green chili paste, oil, dal flour, garlic cloves, mustard, asafoetida, turmeric, jaggery, *etc.*

Method: Sauté the garlic in oil. In a cooker, cook the mango and dal together. Knead the mixture and add dal flour. Add vegetables to the mixture. Then, incorporate crushed chili paste, salt, ground nuts, and jaggery. Cook the mixture until done.

Uses:

1. Culinary: The leaves of *O. corniculata* are edible and can be consumed in salads, soups, and stir-fried dishes. The plant has a tart and tangy taste due to its oxalic acid content, which adds a unique flavor to culinary preparations (Badwaik *et al.*, 2011).

2. Traditional Medicine: In traditional herbal medicine systems, *O. corniculata* has been used for various medicinal purposes. It has shown diuretic, antiscorbutic (vitamin C-rich), and antimicrobial properties. It has been used to treat ailments such as coughs, fevers, gastrointestinal issues, and skin conditions (Badwaik *et al.*, 2011; Sharma *et al.*, 2011).

3. Ornamental Plant: *O. corniculata* is often grown as an ornamental plant in gardens or as a ground cover. It is valued for its attractive trifoliate leaves and delicate yellow flowers, which add aesthetic appeal to landscapes (Sarkar *et al.*, 2020).

CHAPTER 13

Wild Vegetables of the Family Phyllanthaceae

INTRODUCTION

The Phyllanthaceae family, abundant in Southeast Asia and warm regions worldwide, includes 2000 species across 59 genera. These plants vary widely, from tiny herbs to towering trees, with some even being climbers. They can be distinguished by having two ovules per ovary chamber and lacking the milky latex sap found in their close relatives, the Euphorbiaceae. Phyllanthaceae members can be a good source of vitamins like C, A, and B complex, along with minerals like calcium, iron, and potassium (Xu and Deng 2017).

***Phyllanthus amarus* Schum. & Thonn.**

Botanical name: *Phyllanthus amarus*

Family: Phyllanthaceae

Local name: Bhuiavla

Vernacular name:

- **Hindi:** Bhuiavla
- **Bengali:** Bhuiamla
- **Malayalam:** Kilanelli
- **Manipuri:** Chakpa heikru
- **Sanskrit:** Bahupatra
- **Kannada:** Kiru Nelli
- **Telugu:** Nela usiri

Season: June to October

Parts used: Leaves and fruits

Characteristics (Ghosh *et al.*, 2022):

Ganesh Chandrakant Nikalje, Apurva Chonde, Sudhakar Srivastava & Penna Suprasanna

1. Plant Morphology: *P. amarus* is a small, erect annual herb that typically grows up to 30-60 cm in height. It has a slender and branched stem with a green coloration (Fig. **36.1**).

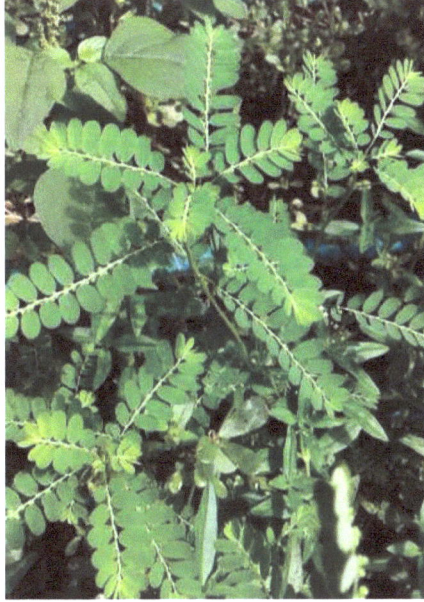

Fig. (36.1). Whole plant of *P. amarus* (PC: Apurva Shankar Chonde).

2. Leaves: The leaves of *P. amarus* are small and alternate in arrangement. They are oblong or elliptical in shape and have a smooth margin. The leaves are light green and possess a glossy texture.

3. Flowers: The flowers of *P. amarus* are small, inconspicuous, and greenish-yellow in color. They are usually unisexual and have a cluster-like arrangement in the leaf axils or at the ends of the branches.

4. Fruits: *P. amarus* produces small, round-shaped fruits that are typically green or yellow in color when ripe. The fruits contain seeds and have a smooth texture.

5. Roots: The roots of *P. amarus* are shallow and fibrous.

Distribution: Asia, India, China, Thailand, Malaysia, Indonesia, Sri Lanka, and the Philippines, Africa- Nigeria, Ghana, South Africa, Kenya, and Tanzania. Australia, America- Brazil, Colombia, Mexico, and the Caribbean, Europe- Spain and Portugal.

Propagation: Seeds

Chemical constituents:

Alkaloids, flavonoids, hydrolysable tannins (ellagitannins), lignans, polyphenols, sterols, and triterpenes (Patel *et al*, 2011).

Recipe:

Ingredients: Younger leaves, finely chopped onion, 5-6 garlic cloves, salt, green chilies, curry leaves, Asafoetida, mustard seeds, turmeric, gram dal, *etc.*

Method: Peel the stalks of the leaves, wash the leaves, and chop them finely. Heat oil in a pan and fry asafoetida, mustard seeds, curry leaves, crushed garlic cloves, and finely chopped onion. When the onion turns slightly yellow, add the chopped vegetables, chili, turmeric powder, soaked dal, and salt to taste. Lightly beat water, cover, and cook in a cooker for 20 minutes.

Uses:

1. Liver Health: *P. amarus* is widely used to support liver health and promote liver detoxification. It is believed to have hepatoprotective properties and may help in managing liver disorders like hepatitis, jaundice, and liver damage caused by toxins (Ogunmoyole *et al.*, 2020; Tiwari, 2021).

2. Kidney Health: *P. amarus* is used to support kidney health and urinary tract function. It is believed to have diuretic properties and may help in treating urinary tract infections, kidney stones, and other kidney-related conditions (Foo *et al.*, 1992; Cartaxo *et al.*, 2010; Ogunmoyole *et al.*, 2020).

3. Digestive Health: It is traditionally used to promote digestive health and alleviate digestive disorders such as indigestion, acidity, and constipation. *P. amarus* is believed to have gastroprotective and anti-inflammatory properties (Mahishi *et al.*, 2005; Shanmugam *et al.*, 2009; Upadhyay *et al.*, 2010).

4. Antiviral and Antimicrobial Activity: *P. amarus* exhibits antiviral and antimicrobial properties and is used to combat viral and bacterial infections. It may be used to support the immune system and help in managing conditions like colds, flu, and respiratory infections (Calixto *et al.*, 1998; Demain *et al.*, 2000; Kassuya *et al.*, 2005; Ramandeep *et al.* 2017).

5. Anti-inflammatory and Antioxidant Properties: It is believed that *P. amarus* possesses anti-inflammatory and antioxidant properties, which may help in reducing inflammation and oxidative stress in the body (Fauré *et al.*, 1990; Demain *et al.*, 2000; Kassuya *et al.*, 2005; Biswas *et al.*, 2020).

6. Diabetes Management: *P. amarus* is used in traditional medicine for managing diabetes. It is believed to have blood sugar-regulating properties and may help in controlling blood glucose levels (Mahishi *et al.*, 2005; Shanmugam *et al.*, 2009).

7. Skin Health: It is used in topical applications for various skin conditions like eczema, rashes, and wounds. *P. amarus* may help in soothing skin irritations and promoting wound healing (Tirimana, 1987; Heyde, 1990).

Phyllanthus emblica L.

Botanical name: *Phyllanthus emblica*

Family: Phyllanthaceae

Local name: Aavala

Vernacular name:

- **Assamese:** Amlaki
- **Bengali:** Amlaki
- **English:** Amla
- **Gujarati:** Amla
- **Hindi:** Amla
- **Kannada:** Aamalaka
- **Malayalam:** Nelli
- **Oriya:** Aula
- **Sanskrit:** Brahmavriksh
- **Telugu:** Amalakamu

Season: Flowering: March-May. Fruiting: June-September

Parts used: Bark, Fruit, Seed

Characteristics (Prananda *et al.*, 2023):

1. Size and Growth Habit: *P. emblica* is a medium-sized tree that typically grows to a height of 8-18 meters (26-59 feet). It has a spreading and dense crown.

2. Leaves: The leaves of *P. emblica* are simple, alternate, and closely set on the branches. They are pinnate or sub-pinnate, meaning they have multiple pairs of leaflets attached to a central axis. The leaflets are small, narrow, and oblong in shape, with a glossy green color.

3. Flowers: The flowers of *P. emblica* are small, greenish-yellow, and inconspicuous. They are arranged in axillary clusters or fascicles.

4. Fruits: The fruits of *P. emblica* are round or spherical berries, which are the key characteristic of the plant. They are green when immature and turn yellowish-green or yellow when ripe. The fruit has a smooth and hard surface with six vertical grooves or ridges (Fig. **36.2**).

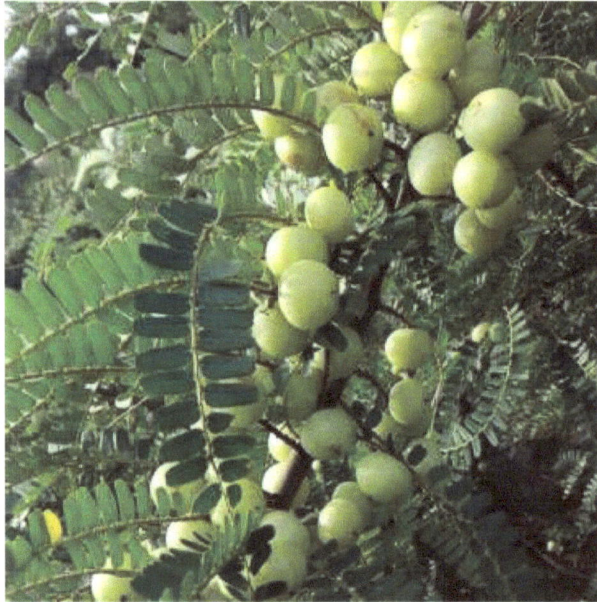

Fig. (36.2). Fruits of *P. emblica* (PC: Apurva Shankar Chonde).

5. Bark: The bark of *P. emblica* is light brown or grayish-brown and has a rough texture. It may have shallow vertical fissures or cracks.

6. Branches and Twigs: The branches of *P. emblica* are slender and often drooping. The twigs are smooth and glabrous (without hairs).

7. Roots: The roots of *P. emblica* are fibrous and shallow, spreading horizontally near the surface of the soil.

Distribution:

It is native to the Indian subcontinent. It is found in various regions of India, including the foothills of the Himalayas, as well as in neighboring countries such as Nepal, Pakistan, Sri Lanka, and Bangladesh.

Propagation: Seeds

Chemical constituents:

Phenols- corilagin, ellagic acid, gallic acid, kaempferol, geraniin, furosin, gallotanins, emblicanins, quercetin, flavonoids, glycosides, and proanthocyanidins (Nisha *et al*, 2004; Kumaran *et al.*, 2006).

Recipe:

Ingredients: aavala, gram flour, mustard seeds, methi seeds, dried red chilies, soap, turmeric powder, cumin-coriander powder, salt, red chili powder.

Method: Boil the aavala (Indian gooseberry), smash it, and remove the seeds. Heat oil in a pan and add mustard seeds, dry red chilies, and fenugreek seeds. After 2 minutes, add turmeric powder, coriander-cumin powder, red chili powder, and salt. Mix well and cook for 1-2 minutes. Then, add the smashed aavala and jaggery. Cook it for 4-5 minutes and serve with rice.

Uses:

1. Antioxidant activity: *P. emblica* is a potent antioxidant, which helps in neutralizing harmful free radicals in the body and protects against oxidative stress and damage (Habib Ur *et al.*, 2007; El-Desouky *et al.*, 2008; Mirunalini *et al.*, 2010).

2. Immune system support: Amla is known to enhance the immune system and improve overall immunity. It helps in fighting off infections, colds, and flu (Duan *et al.*, 2005; Madhuri *et al.*, 2011; Kumar *et al.*, 2012b).

3. Digestive health: It promotes healthy digestion and aids in the absorption of nutrients. Amla is commonly used to alleviate digestive disorders such as acidity, constipation, and indigestion.

4. Hair health: *P. emblica* is beneficial for maintaining healthy hair. It strengthens the hair follicles, prevents hair loss, and promotes hair growth. Amla oil is often used for nourishing the scalp and improving hair texture (Thakur *et al.*, 1989; Kumar *et al.*, 2012b).

5. Skincare: Amla is known for its rejuvenating and anti-aging properties. It helps in improving skin texture, reducing blemishes, and promoting a healthy complexion. Amla extracts or powders are used in various skincare products (Williamson, 2002; Chaudhuri and Ratan, 2004; Fujii *et al.*, 2008).

6. Liver health: *P. emblica* supports liver function and aids in detoxification. It helps in protecting the liver against damage caused by toxins and promotes its proper functioning (Jose and Kuttan, 2000; Tasduq *et al.*, 2005).

7. Heart health: Amla has cardio-protective properties and helps in maintaining healthy heart function. It helps in reducing cholesterol levels, managing blood pressure, and preventing the formation of plaque in the arteries (Prince *et al.*, 2009; Padma *et al.*, 2011; Nabavi *et al.*, 2012).

8. Eye health: *P. emblica* is beneficial for maintaining good eye health. It contains antioxidants that protect the eyes from oxidative stress and may help in preventing age-related macular degeneration and cataracts (Biswas *et al.*, 2001; Shah, 2017).

CHAPTER 14

Wild Vegetables of the Family Plantaginaceae

INTRODUCTION

The Plantaginaceae family, or plantain family, includes over 1,900 species across about 90 genera. Found worldwide, especially in temperate zones, these plants range from common herbs like plantains and foxgloves to ornamental snapdragons. They lack vertical partitions in leaf hairs, have variable flower structures, and typically spiral or opposite leaves. Some members, like Foxglove, are used in medicine for heart treatments (Hamed *et al.*, 2014).

Bacopa monnieri (L.) Wettst.

Botanical name: *Bacopa monnieri*

Family: Plantaginaceae

Local name: Brahmi

Vernacular name:

- **Assamese:** Brahmi
- **Bengali:** Brahmisaka
- **Gujarati:** Baam, Brahmi, Jalanevari, Kadavi luni
- **Hindi:** Baam, Brahmi, Jalbuti, Jalnim, Nirbrahmi, Safed chamani
- **Kannada:** Brahmi, Jala brahmi, Niru brahmi
- **Konkani:** Brahmi
- **Malayalam:** Brahmi
- **Manipuri:** Brahmi-sak
- **Marathi:** Brahmi, Jalabrahmi, Nir brahmi
- **Oriya:** Brahmi, Prusni parnni
- **Sanskrit:** Brahmi, Tiktalonika
- **Tamil:** Nir-p-pirami, Piramiyam, Taray
- **Telugu:** Sambrani aku

Season: Flowering: May-October

Parts used: Whole plant, leaves

Characteristics (Khan *et al.*, 2021):

1. Plant Morphology: It is a small, prostrate, or creeping herb that typically grows in moist or wet environments. It has succulent stems with opposite leaves that are sessile (without stalks). The leaves are small, fleshy, and arranged in pairs along the stem. They are oblong or obovate in shape, with rounded or slightly notched tips (Fig. **37.1**).

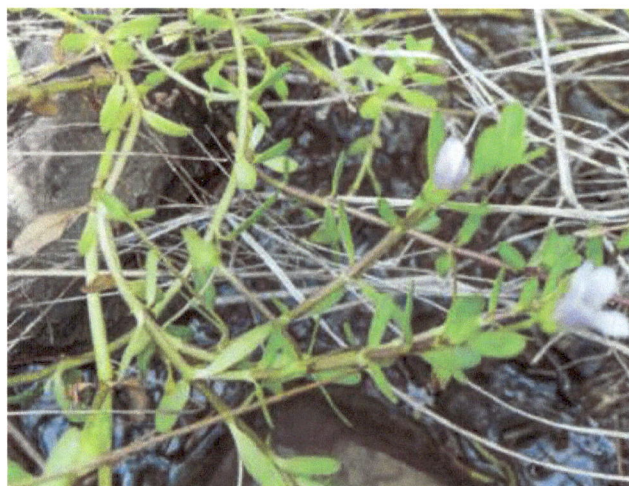

Fig. (37.1). Flowering twig of *B. monnieri* (PC: Shivshankar Chapule).

2. Flowers: It produces small, white, or bluish-purple flowers that are solitary or arranged in clusters in the leaf axils. The flowers are typically five-petaled and have a tubular shape. They bloom throughout the year in favorable conditions.

3. Fruits: After flowering, the plant develops small, round, and capsule-like fruits. These fruits contain tiny, ovoid-shaped seeds.

Distribution: It has a wide geographical distribution. It is native to the wetlands of India, Nepal, Sri Lanka, China, Taiwan, and Vietnam. However, it has been introduced and cultivated in various other regions around the world.

Propagation: Seeds and stem cuttings

Chemical constituents:

Bacosides, alkaloids- brahmine, herpestine, and nicotine, flavonoids- apigenin, luteolin, and quercetin and sterols- stigmasterol, beta-sitosterol, daucosterol, *etc.*

Recipe:

Ingredients: Crushed peanuts, leaves, mustard, cumin, green or red chilies, turmeric powder, hing, garlic.

Method: Use leaves, mainly the central ones from mature leaves. Wash and chop the leaves. Heat oil, and add a small amount of mustard and cumin seeds. Follow with hing or garlic, depending on your preference, and turmeric. Add green or red chilies. Next, introduce the leaves and crushed peanuts. Season with salt to taste. Cook until tender.

Uses:

1. Cognitive Enhancement and Memory Improvement: It has been traditionally used as a memory-enhancing herb and for its potential cognitive benefits. It may support learning, memory retention, and overall cognitive function (Aguiar and Borowski, 2013).

2. Anxiety and Stress Relief: It has been used traditionally as an adaptogen, helping the body and mind cope with stress and anxiety. It may have an anxiolytic effect, promoting a sense of calm and relaxation (Bhattacharya *et al.*, 2000).

3. Neuroprotection: It possesses neuroprotective properties, which may help protect against oxidative stress and neuronal damage. It may have antioxidant effects and can support the health and function of the nervous system (Sairam *et al.*, 2002).

4. Anti-inflammatory and Antioxidant Effects: It contains compounds with anti-inflammatory and antioxidant properties, which may help reduce inflammation and oxidative stress in the body (Sairam *et al.*, 2002).

5. Anti-Epileptic Properties: It has been used traditionally for its potential anti-epileptic effects. It possesses anticonvulsant properties, although further research is needed to understand its mechanisms of action (Sairam *et al.*, 2002).

CHAPTER 15

Wild Vegetables of the Family Plumbaginaceae

INTRODUCTION

The Plumbaginaceae family, or leadwort family, includes about 650 species across 27 genera. Found worldwide, especially in salty environments like coasts and steppes, these plants are known for their vibrant pink and blue blooms. They feature chalk glands to expel excess salt, perennial herbaceous forms with basal leaves, and five-petaled flowers. While some have medicinal uses, their primary value is ornamental, with thrift (Armeria) being a popular example (Kubitzki 1993).

Plumbago zeylanica L.

Botanical name: *Plumbago zeylanica*

Family: Plumbaginaceae

Local name: Chitrak

Vernacular name:

- **Common name:** Chitrak, Plumbago, White leadwort
- **Hindi:** Chitrak
- **Assamese:** Boga agechita
- **Manipuri:** Telhidak angouba
- **Tamil:** chittiramoolam Karimai
- **Malayalam:** Vellakoduveli
- **Kannada:** Chitramula, Chitramulike, Chitraka, Vyaala, Agni
- **Bengali:** Safaid-sitarak
- **Oriya:** Ogni

Season: December to March

Parts used: Roots, leaves and milky juice

Characteristics:

1. Growth Habit: *P. zeylanica* is a perennial herbaceous plant that can grow as a subshrub or groundcover. It has a sprawling or scrambling growth habit, spreading horizontally along the ground or climbing with the help of nearby support (Kapoor, 1990; Roy *et al.*, 2017).

2. Leaves: The leaves of *P. zeylanica* are simple, alternate, and elongated, with an oblong to lanceolate shape. They are smooth-edged, have a glossy texture, and are arranged spirally along the stem (Kapoor, 1990; Roy *et al.*, 2017).

3. Inflorescence: The flowers of *P. zeylanica* are arranged in loose, elongated racemes that emerge from the leaf axils or at the tips of the branches. The racemes can be several inches long and bear multiple flowers (Kapoor, 1990; Roy *et al.*, 2017).

4. Flowers: The plant produces small, tubular-shaped flowers that are typically white or pale blue in color. The flowers are borne in terminal clusters or spikes known as racemes. Each flower has five petals and a slender tube-like corolla (Kapoor, 1990; Roy *et al.*, 2017) (Fig. **38.1**).

Fig. (38.1). Flowering twig of *P. zeylanica* (PC: Shivshankar Chapule).

5. Fruits: After the flowers are pollinated, *P. zeylanica* forms small, rounded capsules that contain multiple seeds. The capsules are green initially but turn

brown as they mature. When mature, they split open to release the seeds (Kapoor, 1990; Roy *et al.*, 2017).

6. Stem and Branches: The stems of *P. zeylanica* are slender, wiry, and often reddish-brown in color. They may have a slightly zigzag pattern and are typically covered in smooth bark (Kapoor, 1990; Roy *et al.*, 2017).

Distribution: South Asia- India, Sri Lanka, Nepal, Bangladesh, and Myanmar, Southeast Asia- Thailand, Laos, Cambodia, and Vietnam, Africa- South Africa, Tanzania, Kenya, and Uganda.

Propagation: Seeds, cuttings, division of older plants.

Chemical constituents:

Plumbagin, coumarins- scopoletin and umbelliferone, flavonoids- apigenin, luteolin, quercetin, and kaempferol derivatives, alkaloids- plumbaginol and plumbaginone, triterpenoids- ursolic acid and oleanolic acid, essential oils containing- limonene, linalool, terpinen-4-ol, and α-pinene

Recipe:

Ingredients: Moong dal, leaves, mustard, cumin, green or red chilies, turmeric powder, hing, garlic, *etc.*

Method: Cook with soaked moong dal. Use only the leaves and remove tough veins, especially the central ones from mature leaves. Wash and chop the leaves. Heat oil and add a small amount of mustard and cumin seeds, followed by hing (or garlic, depending on preference) and green or red chilies. Add the soaked dal (drain it first) and then the leaves, along with turmeric and salt to taste. Cook until tender.

Uses:

1. Digestive Disorders: *P. zeylanica* is traditionally used for digestive issues such as indigestion, flatulence, and stomachaches. It possesses carminative properties that help alleviate these conditions (Shukla *et al.*, 2021).

2. Respiratory Conditions: The plant is used in traditional medicine to treat respiratory ailments like coughs, bronchitis, and asthma. It showed expectorant properties, helping to loosen phlegm and clear the airways (Sheeja *et al.*, 2010; Aleem, 2020).

3. Anti-inflammatory and Analgesic Effects: *P. zeylanica* is known for its anti-inflammatory properties and is used to alleviate pain and inflammation associated with conditions like arthritis and rheumatism (Arunachalam *et al.*, 2010; Sheeja *et al.*, 2010; Aleem, 2020).

4. Skin Disorders: It is used topically in the form of poultices or ointments to treat skin conditions such as eczema, itching, and wounds. It has antiseptic and wound-healing properties (Ghazanfar, 1994; Vishnukanta *et al.*, 2011).

5. In some traditional practices, *P. zeylanica* is used in rituals and ceremonies as it possesses protective and purifying properties (Kumar *et al.*, 2009).

6. It is also sometimes used as an insect repellent or in insecticidal preparations (Kumar *et al.*, 2009).

CHAPTER 16

Wild Vegetables of the Family Poaceae

INTRODUCTION

The Poaceae family, or grasses, comprises over 10,000 species in roughly 700 genera. Dominating grasslands and meadows worldwide, they are essential to many ecosystems. Recognizable by their hollow, linear leaves with parallel veins and spikelet-clustered flowers, Poaceae lack true petals and rely on wind for pollination. This family is vital for food staples like wheat, rice, and corn, and for livestock forage. While some grasses have medicinal uses, their primary value lies in their nutritional content, offering complex carbohydrates, vitamins, and minerals, though some contain anti-nutrients like phytic acid (Khan *et al.*, 2019).

Bambusa arundinacea (Retz.) Willd.

Botanical name: *Bambusa arundinacea*

Family: Poaceae

Local name: Bamboo, Kalaka, Maanga, Baamboo, Velu

Vernacular name:

- **Assamese:** Kotoha-banh, Jaati Baansh
- **Bengali:** Bansha
- **English:** Spiny bamboo, Thorny bamboo
- **Gujarati:** Baambu, Vans
- **Hindi:** Baans
- **Kannada:** Bidiru, Vamsha
- **Kashmiri:** Bons, Bains
- **Malayalam:** Illi, Mula
- **Sanskrit:** Amupah, Ardrapatrakah, Vamsh
- **Tamil:** Kulay-munkil, Mungil
- **Telugu:** Bongu-veduru, Vamsamu
- **Urdu:** Baans, Buns

Ganesh Chandrakant Nikalje, Apurva Chonde, Sudhakar Srivastava & Penna Suprasanna

Season: Flowering and Fruiting: Once in a lifetime, often during September – May

Parts used: All parts

Characteristics:

1. Plant Morphology: It is a large, clumping bamboo that can reach heights of 15-30 meters (49-98 feet) or even higher. It has thick, hollow culms (stems) with a diameter of 10-20 centimeters (4-8 inches). The culms are typically green when young, turning yellowish-brown or grayish-brown as they mature (Bole and Pathak, 1988) (Fig. **39.1**).

Fig. (39.1). Stems and leaves of *B. arundinacea* (PC: Apurva Shankar Chonde).

2. Leaf Arrangement: The leaves are alternate and occur in dense clusters at the nodes of the culms. Each leaf is lanceolate or linear in shape and can be 10-30 centimeters (4-12 inches) long and 2-5 centimeters (0.8-2 inches) wide.

3. Inflorescence: It produces inflorescences called panicles, which emerge from the upper portion of the culms. The panicles are large, drooping clusters of spikelets, with each spikelet containing multiple florets.

4. Culm Sheaths: The culm sheaths are leathery and have a dark brown or purplish-brown color. They cover the young culms and gradually fall off as the culms mature.

Distribution:

Native to southern Asia (India, Bangladesh, Sri Lanka and Indochina). It is also naturalized in Seychelles, Central America, the West Indies, Java, Malaysia, Maluku, and the Philippines.

Propagation: Rhizome and culm cuttings.

Chemical constituents:

Phenolic compounds- flavonoids and phenolic acids, lignans, dietary fibers, and essential minerals- calcium, potassium, magnesium, and iron.

Recipe:

Ingredients: Chopped sprouts, onion, soaked lentils or gram, red, chili powder, wet coconut, oil turmeric, salt, mustard, asafoetida, *etc.*

Method: Cook the chopped sprouts in a cooker with three to four whistles. Heat oil in a pan, add mustard seeds, asafoetida and fry. Add chopped onion and fry. Then add cooked sprouts and soaked dal. Then add turmeric powder, red chili powder, and salt and saute well. Steam the vegetables and cook. Then spread the grated wet coconut and make dry vegetables. If you want to make thin vegetables, add a little water to the vegetables. Finely chop the wet coconut and saute the vegetables.

Uses:

1. Nutritional: The shoots are consumed as a source of dietary fiber, vitamins, minerals, and other nutrients. They are valued for their low-calorie content and high nutritional value (Nazreen, 2022).

2. Antioxidant Activity: It has been found to exhibit antioxidant activity, which can help protect the body against oxidative stress and associated diseases. The presence of phenolic compounds and flavonoids contributes to its antioxidant potential (Nazreen, 2022).

3. Anti-inflammatory Properties: Studies have indicated that extracts possess anti-inflammatory properties. These properties make it a potential candidate for the development of anti-inflammatory drugs (Nazreen, 2022).

4. Wound Healing: The extracts have shown potential in promoting wound healing. They exhibit antimicrobial activity and accelerate the wound-healing process by aiding in the formation of new tissue (Nazreen, 2022).

5. Antimicrobial Activity: Certain parts such as leaves and shoots, have been found to possess antimicrobial properties. They have demonstrated inhibitory effects against various pathogenic bacteria and fungi (Nazreen, 2022).

6. It is used in various applications such as making traditional and musical instruments, boat rafts, construction, furniture and flooring, fencing and fodder for cattle, utensils for cooking, and in the management of wastewater (Sharma *et al.*, 1980).

7. The root (burnt root) is applied to ringworm, bleeding gums, and painful joints (Khare, 2007). Seeds are acrid, laxative, and said to be beneficial in strangury and urinary discharges (Chopra *et al.*, 1958). The bark is used for skin eruptions (Khare, 2007). The leaf is an emmenagogue, antileprotic, febrifuge, and bechic, and is used in haemoptysis (Khare, 2007).

Dendrocalamus strictus (Roxb.) Nees

Botanical name: *Dendrocalamus strictus*

Family: Poaceae

Local name: Bans, Velu

Vernacular name:

- **Hindi:** Bans
- **Manipuri:** Khokwa
- **Marathi:** Bans, Udha, Velu
- **Tamil:** Ciru-munkil, Kal-munkil, Kattu-munkil
- **Malayalam:** Kallanmula
- **Telugu:** Potu veduru
- **Kannada:** Bidiru
- **Bengali:** Bamsa, Karali
- **Oriya:** Bans
- **Konkani:** Vaaso
- **Urdu:** Bansa
- **Gujarati:** Bans
- **Sanskrit:** Vansha
- **Nepali:** Kaban bans

Season: Flowering: November to February; fruiting: February to April

Parts used: Leaves, Young shoots, stem

Characteristics:

1. Size: *D. strictus* is a tall and robust bamboo species, capable of reaching heights up to 30 meters (98 feet). It has a thick culm (stem) with a diameter of about 10-20 cm (4-8 inches) (Flora of China Editorial Committee, 2015).

2. Culm Color: The culms of *D. strictus* are typically green in their initial stage but gradually mature to a yellowish-brown or pale green color (Flora of China Editorial Committee, 2015).

3. Internode Structure: The culm is divided into distinct sections called internodes. The internodes of *D. strictus* are typically long and straight, with a smooth surface (Flora of China Editorial Committee, 2015).

4. Node Structure: At the junction of each internode, there is a node. The nodes of *D. strictus* have small raised rings or nodal ridges, which are characteristic of this species (Flora of China Editorial Committee, 2015) (Fig. **39.2**).

Fig. (39.2). Stem of *D. strictus (PC: Sumit Bhosle).

5. Leaves: The leaves of *D. strictus* are long and slender, with a lanceolate shape. They are typically dark green in color and have a pointed tip (Flora of China Editorial Committee, 2015).

6. Rhizome System: *D. strictus* has an extensive rhizome system that spreads underground. The rhizomes are thick and can give rise to new shoots and culms (Flora of China Editorial Committee, 2015).

Distribution:

India- Maharashtra, Madhya Pradesh, Gujarat, Rajasthan, Uttar Pradesh, Bihar, and West Bengal, Nepal, Bangladesh, Bhutan, Myanmar, Sri Lanka.

Propagation: Seed, Rhizome, culm and branch cuttings

Chemical constituents:

Flavonoids, phenolic compounds, lignans, alkaloids, terpenoids, sterols, *etc.*

Recipe:

Ingredients: Chopped sprouts, onion, soaked lentils or gram, red chili, wet coconut, oil turmeric, salt, mustard, asafoetida, *etc.*

Method: Cook the chopped sprouts in a pressure cooker for three to four whistles. Heat oil in a pan and add mustard seeds and asafoetida. Fry them for a while. Then, add chopped onion and continue frying. Next, add the cooked sprouts and soaked dal. Sprinkle turmeric powder, red chili powder, and salt. Saute the mixture well. Steam the vegetables until they are cooked. Finally, garnish with grated fresh coconut to make dry vegetables. If you prefer a thinner consistency, you can add a little water to the vegetables. Finely chop the fresh coconut and saute it along with the vegetables.

Uses:

1. Food and Culinary: Certain parts of *D. strictus*, such as bamboo shoots, are edible and used in various cuisines. Bamboo shoots are known for their tender texture and mild flavor. They are used in stir-fries, soups, curries, and salads, adding a unique taste and texture to the dishes (Muthukumar *et al.*, 2006).

2. Traditional and Cultural Practices: Bamboo holds significant cultural and traditional value in many societies. It is used in the construction of traditional dwellings, religious structures, and ceremonial items. Bamboo plays an important

role in cultural festivals, crafts, and artistic expressions (Muthukumar *et al.*, 2006).

3. Environmental Protection and Landscaping: Bamboo is recognized for its environmental benefits. It helps in soil erosion control, water conservation, and carbon sequestration. It is also used in landscaping for creating natural barriers, privacy screens, and ornamental gardens (Muthukumar *et al.*, 2006).

4. Construction and Building: Bamboo is widely used as a construction material for various purposes such as making scaffolding, temporary structures, and even permanent buildings. Its strength, flexibility, and lightweight properties make it an excellent alternative to traditional construction materials (Muthukumar *et al.*, 2006).

5. Furniture and Handicrafts: Bamboo is extensively used for making furniture and handicrafts. It is known for its durability, aesthetic appeal, and eco-friendliness. Bamboo furniture, including chairs, tables, and shelves, is popular due to its natural beauty and sustainable characteristics (Muthukumar *et al.*, 2006).

6. Paper and Pulp Industry: Bamboo fibers are used in the paper and pulp industry for producing high-quality paper. The long and strong fibers of *D. strictus* make it suitable for manufacturing various paper products, including writing paper, tissue paper, and cardboard (Muthukumar *et al.*, 2006).

7. Musical Instruments: Bamboo is also used in the manufacturing of musical instruments. Bamboo flutes, wind chimes, and percussion instruments are examples of musical instruments made from *D. strictus*. Bamboo's unique resonance and sound qualities make it a favored material for instrument production (Muthukumar *et al.*, 2006).

CHAPTER 17

Wild Vegetables of the Family Polygonaceae

INTRODUCTION

The Polygonaceae family, or knotweed family, includes around 1,200 species across 48 genera. Found globally, especially in temperate zones, they feature swollen stem nodes, alternate leaves with sheath-like stipules, and clustered small flowers. Buckwheat, a nutritious seed and flour source, is a culinary star of the family. Other edible members include rhubarb and sorrel, with some species having medicinal uses. Overall, the family offers nutritional value with seeds rich in protein and fiber, though some members may contain anti-nutrients (Brandbyge 1993).

***Rumex vesicarius* L.**

Botanical name: *Rumex vesicarius*

Family: Polygonaceae

Local name: Aambat chukka

Vernacular name:

- **Hindi:** Chooka, Chukra, Chukrika, Lolika, Shatavedhi
- **Manipuri:** Torong khongchak
- **Tamil:** Cukkan-kirai
- **Tangkhul:** Hangam ashinba
- **Telugu:** Chukka kura, Pulla-prabba
- **Kannada:** Chukki soppu, Chukra
- **Bengali:** Bun palung
- **Nepali:** Amilo bethe, Bhote palungo
- **Urdu:** Tukhm hummaz
- **Assamese:** Chuka sak
- **Sanskrit:** Amlavetasa, Chukram, Chukrika, Lolika, Shatavedhi

Season: Flowering and fruiting: March to May

Ganesh Chandrakant Nikalje, Apurva Chonde, Sudhakar Srivastava & Penna Suprasanna

Parts used: Leaves

Characteristics (Boulos 1999):

1. Leaves: The leaves of *R. vesicarius* are elongated and lanceolate or oblong in shape. They are typically medium to dark green in color and have prominent parallel veins. The leaves are arranged in a basal rosette, meaning they emerge directly from the base of the plant.

2. Stems: The stems of *R. vesicarius* are slender and erect, growing up to about 60 centimeters in height. The stems may have a reddish or purplish tinge and are usually unbranched or sparsely branched.

3. Flowers: *R. vesicarius* produces small, inconspicuous flowers that are arranged in dense clusters called inflorescences. The flowers are typically greenish to reddish-brown in color and lack showy petals. They are wind-pollinated (Fig. **40.1**).

Fig. (40.1). Inflorescence of *R. vesicarius* (PC: Suresh Shingare).

4. Fruits: After flowering, *R. vesicarius* produces small, three-sided fruits called achenes. These fruits are enclosed within reddish or brownish membranous coverings called perianths or tepals. The perianths are often inflated, giving the plant its common name "Bladder Dock."

Distribution:

Africa- Egypt, Libya, Algeria, Morocco, Sudan, Ethiopia, Greece, Cyprus, Turkey, and Israel, also found outside Africa from the Mediterranean east to India.

Propagation: Seeds

Chemical constituents:

Anthraquinones- chrysophanol, emodin, and physcion, tannins, flavonoids, phenolic acids- caffeic acid and ferulic acid.

Recipe:

Ingredients: Ambat Chuka leaves with stems, peanuts soaked in water for 2 hrs, soaked Chana dal, Chana dal flour, jaggery, salt, garlic, cumene, mustard, turmeric powder, oil, *etc.*

Method: Pressure cook the chukka, peanuts, and chana dal together for one whistle. In a wok, heat oil and add mustard seeds. Once they crackle, add cumin seeds, garlic cloves, and green chilies. Add turmeric powder and the cooked chukka mixture. Season with salt and jaggery, and cook for a few minutes. Adjust the water quantity to achieve the desired consistency, usually similar to a dal, which can be eaten with rice. Once the water starts boiling, gradually add besan while stirring continuously. Cook for a few more minutes. Serve hot with bhakri or plain rice.

Uses:

1. Traditional Medicine: In traditional medicinal practices, various parts of *R. vesicarius*, such as leaves, roots, and seeds, have been used for their potential therapeutic properties. It has been traditionally used for treating digestive disorders, including constipation and stomach ailments. The plant is also used as a diuretic and for its potential laxative effects (Boulos, 1983; Litvinenko *et al.*, 2003; Mostafa *et al.*, 2011; Hafaz *et al.*, 2022).

2. Culinary: In some regions, the young leaves and shoots of *R. vesicarius* are consumed as leafy greens in culinary preparations. They can be cooked, stir-fried, or used in salads. The plant has a tangy or sour taste, which adds flavor to dishes (Quézel *et al.*, 1962).

3. Livestock Feed: *R. vesicarius* is sometimes used as fodder for livestock. The plant is rich in nutrients and can be a valuable food source for animals, particularly in areas with limited grazing resources (Mostafa *et al.*, 2011).

CHAPTER 18

Wild Vegetables of the Family Portulacaceae

INTRODUCTION

The Portulacaceae, or purslane family, recently underwent a taxonomic revision. Previously, it included around 20 genera and 500 species; now, it comprises only 1 genus, *Portulaca*, with roughly 115 species. These plants are recognized by their fleshy, often rosette-clustered leaves and showy flowers with five petals. They are rich in vitamins A and C, omega-3 fatty acids, fiber, and minerals. However, some species may contain anti-nutrients like lectins and tannins (Xu and Deng 2017).

Portulaca oleracea L.

Botanical name: *Portulaca oleracea*

Family: Portulacaceae

Local name: Chiu, Ghol

Vernacular name:

- **Assamese**: Hah thegia
- **Bengali**: Nunia Sag
- **English**: Indian parselane
- **Gujarati**: Ghol
- **Hindi**: Lunia
- **Malayalam**: Cheriyagolicheera
- **Kannada**: Doodagooni Soopu
- **Sanskrit**: Brihalloni
- **Tamil**: Karpakantakkirai

Season: All season

Parts used: Leaves

Characteristics:

1. Growth Habit: *P. oleracea* is a low-growing, succulent annual plant. It has a prostrate or spreading growth habit, with stems that radiate from a central point or grow along the ground (Purslane, 1995; Chowdhary *et al.*, 2013).

2. Leaves: The leaves of *P. oleracea* are fleshy, smooth, and alternate in arrangement. They are spoon-shaped or obovate and vary in size, ranging from small to medium-sized. The leaves are typically shiny and may have a reddish or green color, depending on the cultivar and environmental conditions (Purslane, 1995; Chowdhary *et al.*, 2013) (Fig. **41.1**).

Fig. (41.1). Vegetative twig of *P. oleracea* (PC: Apurva Shankar Chonde).

3. Stems: The stems of *P. oleracea* are thick, smooth, and reddish or greenish in color. They are succulent and often form a dense mat-like structure close to the ground (Purslane, 1995; Chowdhary *et al.*, 2013).

4. Flowers: *P. oleracea* produces small, showy flowers that bloom at the tips of the stems. The flowers have five petals and come in a range of colors, including yellow, orange, pink, and white. Each flower is typically about 1 centimeter in diameter (Purslane, 1995; Chowdhary *et al.*, 2013).

5. Reproductive Structures: After flowering, *P. oleracea* develops small, capsule-like fruits that contain numerous tiny black seeds. The fruits are located at the base of the flower petals and eventually split open to release the seeds.

Distribution:

Europe- Greece, Italy, Spain, Portugal, United Kingdom, France, Germany, and Eastern European countries. Asia- India, China, Japan, Korea, and various Southeast Asian countries, Americas- United States, Mexico, Canada, Brazil, Argentina, Chile, and Colombia, Africa- Egypt, Morocco, South Africa, Kenya, Nigeria, and Tanzania, Australia and Oceania- New Zealand, Fiji, Papua New Guinea, and parts of Polynesia.

Propagation: Seeds and stem cuttings.

Chemical constituents:

Omega-3 fatty Acids- alpha-linolenic acid (ALA), Flavonoids- quercetin, kaempferol, and rutin, phenolic compounds- coumarins, caffeic acid, and ferulic acid, Betalains, Vitamins- vitamin C, vitamin A, vitamin E, Minerals- magnesium, calcium, potassium, iron, and polysaccharides.

Recipe:

Ingredients: Younger leaves, finely chopped onion, 5-6 garlic cloves, salt, green chilies, curry leaves, Asafoetida, mustard seeds, turmeric, gram dal, *etc.*

Method: Peel the stems of the leaves, wash the purslane leaves, and chop them finely. Heat oil in a pan and fry asafoetida, mustard seeds, curry leaves, crushed garlic cloves, and finely chopped onion. When the onion turns slightly yellow, add the chopped vegetables, chili, turmeric powder, soaked dal, and salt to taste. Lightly beat water, cover, and cook over low flame.

Uses:

1. Culinary: *P. oleracea* is consumed as a leafy vegetable in various cuisines around the world. The tender leaves and stems are used in salads, stir-fries, soups, and stews. It has a slightly tangy and lemony flavor that adds a refreshing taste to dishes (Rahimi *et al.*, 2019).

2. Nutritional Value: *P. oleracea* is highly nutritious and is rich in essential nutrients. It is a good source of vitamins A, C, and E, as well as minerals such as magnesium, calcium, potassium, and iron. It also contains omega-3 fatty acids, which are beneficial for heart health (Malek *et al.*, 2004; Zhou *et al.*, 2015).

3. Medicinal: *P. oleracea* has been used in traditional medicine for various health conditions. It is expected to have several medicinal properties, including:

Anti-inflammatory: Purslane is known for its anti-inflammatory effects and may help reduce inflammation in the body (Chan *et al.*, 2000).

Antioxidant: It contains flavonoids and other compounds that act as antioxidants, helping to protect against oxidative stress and cellular damage (Zhou *et al.*, 2015).

Diuretic: *P. oleracea* has diuretic properties, which can help promote urine production and flush out toxins from the body (Lee *et al.*, 2012).

Gastrointestinal Health: It has been used to alleviate digestive issues such as indigestion, constipation, and diarrhea (Iranshahy *et al.*, 2017).

4. Skin Care: Purslane has been used in skincare preparations due to its antioxidant and soothing properties. It may help moisturize the skin, reduce inflammation, and promote healing of minor skin irritations (Gai *et al.*, 2016).

5. Animal Feed: *P. oleracea* is sometimes used as fodder for livestock and is appreciated for its high nutritional value (Dkhil *et al.*, 2011).

Portulaca quadrifida L.

Botanical name: *Portulaca quadrifida*

Family: Portulaceae

Local name: Chivli

Vernacular name:

- **Sanskrit:** Paciri, Paviri.
- **English:** Chicken weed, Wild purslane, Single-flowered purslane; Small-leaved purslane; Ten o'clock plant
- **Hindi:** Paviri, Chounlayi, Khate Chawal
- **Telugu:** Goddu pavelli
- **Marathi:** Rangol, Khate chanval
- **Tamil:** Pasalai keerai
- **Malayalam:** Neelakeera
- **Kannada:** Gooni soppu, Hali bachchdi hali dajjili

- **Sindhi:** Lunak
- **Chinese:** Si lie ma chi xian, Si ban ma chi xian
- **French:** Pourpier
- **Burma:** Mya-byit, Mya-byit-gale

Season: October to April

Parts used: Leaves, Seed

Characteristics:

1. Growth Habit: *P. quadrifida* is a low-growing, herbaceous annual plant. It has a prostrate or spreading growth habit, with stems that radiate from a central point or grow along the ground.

2. Leaves: The leaves of *P. quadrifida* are fleshy and succulent. They are alternate in arrangement and typically occur in clusters along the stems. The leaves are small, obovate or spatulate in shape, and have rounded or obtuse tips. Each leaf is about 1-3 centimeters long and may have a reddish or greenish color, depending on environmental conditions.

3. Stems: The stems of *P. quadrifida* are thin, succulent, and often reddish in color. They trail along the ground or form a low mat-like structure.

4. Flowers: *P. quadrifida* produces small, showy flowers that bloom at the tips of the stems. The flowers have four petals and come in various colors, including yellow, pink, or white. Each flower is typically about 1 centimeter in diameter (Fig. **41.2**).

5. Reproductive Structures: After flowering, *P. quadrifida* develops small, capsule-like fruits that contain numerous tiny black seeds. The fruits are located at the base of the flower petals and eventually split open to release the seeds.

Fig. (41.2). Flowering twig of *P. quadrifida* (PC: Rajesh Ghumatkar).

Distribution:

China, India, Nepal, Pakistan; Africa; North America: Caribbean Islands.

Propagation: Seed

Chemical constituents:

Omega-3 fatty acids- alpha-linolenic acid (ALA), flavonoids, phenolic compounds, betalains- betacyanins and betaxanthins, polysaccharides (Das *et al.*, 2013).

Recipe:

Ingredients: Chivli vegetables, dal flour, oil, frying ingredients, salt, garlic, green chilies, *etc.*

Method: Select Chivli vegetables carefully, wash them thoroughly, and chop them finely. Heat oil in a pan and fry. Cut the chilies into pieces and add them to the mixture. Then add the chopped vegetables. Add salt to taste and sauté for 2 minutes. This vegetable cooks quickly. Sprinkle the dal flour slowly over it, ensuring there are no lumps. Move the mixture to one side. This vegetable cooks quickly and has a light and loose texture. Sometimes, crushed garlic cloves can be added for enhanced flavor.

Uses:

1. Culinary: *P. quadrifida* is sometimes consumed as a leafy vegetable in certain tribal regions. The tender leaves and stems can be added to salads, soups, stir-fries, and other culinary preparations. (Grubben *et al.*, 2004; Tariku *et al.*, 2017; Verma *et al.*, 2017).

2. Traditional Medicine: In some traditional medicine systems, *P. quadrifida* has been used for various medicinal purposes. It has been used to alleviate conditions such as fever, digestive disorders, and inflammatory conditions (Mulla *et al.*, 2010; Das *et al.*, 2013).

3. Animal Feed: *P. quadrifida* can serve as a source of forage for livestock. It is sometimes used as fodder due to its high nutritional content and availability (Das *et al.*, 2013).

4. Soil Erosion Control: *P. quadrifida* has been used in some areas for its ability to control soil erosion. Its spreading growth habit and extensive root system can help stabilize soil in certain landscapes (Das *et al.*, 2013).

5. Ornamental Plant: Some gardeners and horticulture enthusiasts cultivate *P. quadrifida* as an ornamental plant. The attractive foliage, vibrant flowers, and low-growing habit make it suitable for ground covers, hanging baskets, or container gardening (Das *et al.*, 2013).

CHAPTER 19

Wild Vegetables of the Family Rubiaceae

INTRODUCTION

The Rubiaceae family, or coffee family, includes over 13000 species in about 630 genera, mainly found in the tropics. They are recognizable by their simple, opposite leaves with stipules at the base and symmetrical flowers. Economically important members include coffee and madder, with additional medicinal uses and ornamental species. While offering a range of vitamins and beneficial compounds, some members may contain antinutrients (Xu and Chang 2017).

Meyna laxiflora Robyns

Botanical name: *Meyna laxiflora*

Family: Rubiaceae

Local name: Alu, aliv

Vernacular name:

- **Common name:** Muyna
- **Hindi:** Muyna, Pundrika
- **Manipuri:** Heibi
- **Marathi:** Huloo, Alu
- **Tamil:** Manakkarai
- **Telugu:** Visikilamu, Chegagadda
- **Kannada:** Chegugadde, Achura mullu, Mullakare, Gobergally
- **Oriya:** Gurbeli
- **Konkani:** Helu
- **Sanskrit:** Madan, Pindituka
- **Mizo:** Mawntawrawkawk

Season: June to September

Parts used: Fruits, leaves.

Ganesh Chandrakant Nikalje, Apurva Chonde, Sudhakar Srivastava & Penna Suprasanna

Characteristics (Rymbai *et al.*, 2022):

1. Habitat: *M. laxiflora* is commonly found in tropical and subtropical forests, displaying a particular affinity for open areas or edges. This adaptable plant species demonstrates tolerance to a diverse range of soil types and light conditions.

2. Habit: Exhibiting the characteristics of a small tree or large shrub, *M. laxiflora* typically attains a height ranging from 3 to 6 meters. Notably, it is armed with straight, paired spines measuring 1.5-2 cm in length.

3. Leaves: The leaves of *M. laxiflora* are arranged either oppositely or in whorls of three. They take on an elliptic-oblong to ovate-lanceolate shape, measuring 3.5-15 x 1.2-10 cm. Both surfaces of the leaves are glossy and devoid of hair. The base of the leaf is either round or acute, while the apex is acuminate. Prominent midribs and fine lateral veins contribute to the leaf's overall appearance. Stipules, which are triangular and interpetiolar, are also present.

4. Flowers: The flowers of *M. laxiflora* are small and greenish-white, arranged in cymes on leafless nodes. Both the calyx and corolla exhibit 4-5 lobes each. The corolla is short and tubular, with stamens inserted at the mouth. The ovary is superior and 2-celled.

5. Fruits: The fruits of *M. laxiflora* are nearly globose fleshy drupes. When ripe, they appear smooth and take on a yellow hue, measuring about 1-2 cm in diameter. Each fruit contains 4-5 pyrenes, with each pyrene housing a single seed (Fig. **42.1**).

Fig. (42.1). Fruits of *M. laxiflora* (PC: Apurva Shankar Chonde).

Distribution:

Widespread in India, Nepal, Myanmar, and northern Thailand.

Propagation: Seed

Chemical constituents:

Carbohydrates, starch, proteins, tannins, saponins and alkaloids, flavonoids, steroids, proteins and amino acids (Janarthanan *et al.*, 2018).

Recipe:

Ingredients: Fruits, finely chopped onion, 5-6 garlic cloves, salt, green chilies, and curry leaves, Asafoetida, grated coconut, mustard seeds, turmeric, *etc.*

Method: Chop the fruits. Heat oil in the pan and fry asafoetida, mustard seed, curry leaves, crushed garlic cloves, and finely chopped onion. When the onion turns slightly yellow, add the chopped vegetables, chili, turmeric powder, and salt to taste; lightly beat water and cover, and cook over low flame. Once the vegetables are cooked, add wet coconut on top.

Uses:

1. Food: When ripe, the fruits of *M. laxiflora* are palatable when consumed fresh, offering a delicately sweet and mildly acidic taste. In certain tribal communities, young leaves are culinary delights, commonly cooked and consumed as a vegetable. The fruits can undergo fermentation to craft a distinctive wine, distinguished by its unique color and aroma (Deshmukh *et al.*, 2011).

2. Medicinal: *M. laxiflora* has been utilized by indigenous communities for centuries for medicinal purposes. It has been employed to address various ailments, including stomach aches and diarrhoea (Quazi *et al.*, 2014; Dhodade *et al.*, 2019). Additionally, it has been utilized for managing menstrual problems and urinary tract infections. The plant has proven effective in treating skin infections and wounds. Traditional medicine incorporates *M. laxiflora* for diabetes management and blood sugar control. It also exhibits anti-inflammatory properties and provides relief from pain (Patil *et al.*, 2005; Burahohain, 2008; Kamble *et al.*, 2009; Deshmukh *et al.*, 2011).

3. Other:

Construction: The wood of *M. laxiflora* is known for its strength and durability, making it a suitable material for constructing poles, furniture, and tool handles (Patil *et al.*, 2005.)

Fencing: The plant's spines find application in constructing fences, serving as an effective deterrent against both livestock and intruders (Patil *et al.*, 2005).

Fuel: The wood, owing to its combustible properties, serves as a valuable resource for firewood and charcoal production (Patil *et al.*, 2005).

Ornamental: *M. laxiflora* can be cultivated for ornamental purposes, appreciated for its attractive foliage and flowers, enhancing the aesthetic appeal of landscapes (Patil *et al.*, 2005).

Tamilnadia uliginosa (Retz.) Tirveng. & Sastre

Botanical name: *Tamilnadia uliginosa*

Family: Rubiaceae

Local name: Pendhur, Pendari

Vernacular name:

- **Assamese:** Bakhar bengena, Bana bengena
- **Bengali:** Kusum
- **Coorgi:** Kare mara
- **Garo:** Agendra, Suskeng
- **Kannada:** Kare
- **Konkani:** Pendari
- **Malayalam:** Malankaara, Pinticcakka
- **Marathi:** Pendari
- **Oriya:** Potua, Tela korda
- **Sanskrit:** Pindalu, Pinditaka
- **Tamil:** Peru-n-karai
- **Telugu:** Adavi manga

Season: Flowering and fruiting: April-June

Parts used: fruit

Characteristics (Bagga, 2018):

1. Leaves of *T. uliginosa* are arranged simply, opposite, and in a decussate pattern. The stipules are interpetiolar, broadly triangular with acuminate tips, measuring 3-5 x 5-7 mm. The petiole is stout and glabrous, typically 5-10 mm long. The lamina is obovate, oblanceolate, or obovate-oblong, ranging from 5-18 cm in length and 2-8 cm in width. The base is cuneate or attenuate (tapering), and the apex is obtuse (blunt) or round. The margin is entire (smooth without teeth), and the surface is glabrous (smooth) above and pubescent (hairy) and glaucous (whitish-blue) beneath. Venation is characterized by prominent lateral nerves, 5-10 pairs, arranged in a pinnate (feather-like) manner. Intercostae is reticulate (net-like) and slender, with the presence of domatia (structures providing shelter for mites).

2. The stem is terete (round) and pubescent.

3. The flowers of *T. uliginosa* are bisexual, white, and possess a strong fragrance. They are approximately 5 cm across and exhibit dimorphism. Large flowers are sessile (stalkless) and solitary at the end of branchlets, while small flowers are stalked. The calyx is tube-shaped, silky pubescent at the throat, with 5 suborbicular lobes. The corolla has a short tube with 5 large, spreading, orbicular lobes, and is hairy at the mouth. There are 5 stamens with linear anthers. The ovary is inferior, 2-celled, with many ovules. The style is stout, and the stigma is thick and 2-lobed.

4. The fruit is a berry, ovoid or ellipsoid in shape, measuring around 4-5 cm long and 2.5-3.5 cm wide. The surface is smooth and yellow. Seeds are smooth and compressed in shape (Fig. **42.2**).

Fig. (42.2). Fruits of *T. uliginosa* (PC: Apurva Shankar Chonde).

Distribution:

It is found in the forests of Sri Lanka, India, Pakistan, Myanmar, Gujarat, Maharashtra, Karnataka, Kerala, Tamil Nadu, and Assam.

Propagation: Seed

Chemical constituents:

Alkaloids, coumarin, cardiac glycosides, glycosides, phenol, saponin and tannin.

Recipe:

Ingredients: Fruits, red chili powder, salt, turmeric, oil, cumin, mustard, sugar, black sesame paste, *etc.*

Method: Raw fruits should be washed and steamed in a cooker. Then peel them like boiled potatoes and remove the seeds. Heat oil in a pan and add cumin seeds and mustard seeds. Then put boiled strawberry pieces on it. Then add red chili powder, salt, turmeric powder, a little sugar, and black sesame seeds and steam the vegetables.

Uses:

1. In traditional medicine: The roots of *T. uliginosa* are known for their bark, which possesses cooling and diuretic properties. The bark is commonly used to treat conditions such as diarrhoea, dysentery, and biliousness. Additionally, the ground roots are applied externally to address skin issues such as acne, pimples, and wounds (Deepthy *et al.*, 2016).

2. The leaves of *T. uliginosa* play a role in traditional medicine as well. A decoction made from the leaves is used to treat continuous sneezing and reduce joint pain. The juice extracted from the leaves is often combined with other herbs to create remedies for conditions like coughs and urinary tract infections (Deepthy *et al.*, 2016).

3. The unripe fruits of this plant are utilized as fish poison, and when boiled and mixed with salt and chili peppers, they become a consumable remedy for diarrhoea and dysentery relief (Bagga, 2018).

4. In terms of its use as food, the ripe fruits of *T. uliginosa* are eaten raw in certain regions. Young fruits and flowers are also consumed in specific areas (Watt *et al.*, 1972). Additionally, the leaves are boiled and consumed as a vegetable.

5. Beyond medicinal and food applications, *T. uliginosa* serves other purposes. The leaves and fruits are utilized as fodder for deer and cattle. The unripe fruits are employed as a color intensifier in dyeing processes. Moreover, the wood of the plant is used as firewood in some regions.

CHAPTER 20

Wild Vegetables of the Family Sapindaceae

INTRODUCTION

The Sapindaceae family, or soapberry family, includes over 1900 species across 144 genera, primarily in tropical and subtropical regions. These plants, ranging from trees to vines, are identified by their compound leaves and typically unsymmetrical flowers. The family includes both sweet fruits like lychee and rambutan and sources of maple syrup. Some species have medicinal uses, but soapberries contain saponins, which can be irritating. While offering vitamins and minerals in some fruits, some members may also contain anti-nutrients (Buerki *et al.*, 2021).

Cardiospermum halicacabum L.

Botanical name: *Cardiospermum halicacabum*

Family: Sapindaceae

Local name: Kapalphodi

Vernacular name:

- **Assamese:** Kapaal phuta lata
- **Hindi:** Kanphuta, Kapalphodi
- **Kannada:** Agniballi, Bekkina budde gida, Bekkina toddina balli, Chitaki hambu, Erumballi, Jotishmati, Kanakaaya, Minchuballi, Kangunge, Kangonge, Buddakaakarateege
- **Konkani:** Kanphuti, Kapala phodi
- **Malayalam:** Jyotishmati, Karuttakunni, Paluruvam, Uzhinja
- **Manipuri:** Poklaobi
- **Nepali:** Jyotishmati, Kapaal phodi, Kesh lahara
- **Odia:** Jyotishmati, Phutu phutuka
- **Sanskrit:** Karnasphota, Sphutavalkali

- **Tamil**: Korravan, Mutakkorran
- **Telugu**: Buddakakara, Jyotishmati

Season: Flowering and Fruiting: June-November.

Parts used: Fruits, leaves and seeds

Characteristics:

1. Vine-like plant: *C. halicacabum* is a climbing or trailing plant that grows as a vine. It can reach lengths of several meters, with the stem twisting around nearby support structures (Fig. **43.1**).

Fig. (43.1). Vine of *C. halicacabum* (PC: Sumaiya Siddiqui).

2. Leaves: The leaves are compound, meaning they are composed of multiple leaflets. Each leaf typically consists of three leaflets, arranged in an alternate pattern along the stem. The leaflets are ovate or lanceolate in shape, with serrated margins (Duke *et al.*, 1985; Senthilkumar *et al.*, 2013).

3. Flowers: The flowers are small and white or pale yellow in color. They are arranged in loose clusters or racemes. Each flower has five petals and a prominent

central column (Duke *et al.*, 1985; Senthilkumar *et al.*, 2013).

4. Fruits: The characteristic feature of *C. halicacabum* is its unique fruit, which is a small, inflated capsule resembling a balloon. The fruit is initially green but turns brown as it matures. When mature, the fruit dries up and splits open, revealing three black seeds with a white, heart-shaped mark on one side. This heart-shaped mark gives the plant its common name, "cardio" meaning heart, and "spermum" referring to the seeds (Senthilkumar *et al.*, 2013).

Distribution:

India: Assam, Bihar, Gujarat, Jammu & Kashmir, Maharashtra, Manipur, Kerala, Odisha, Punjab, Rajasthan, Tamil Nadu; South America.

Propagation: Seed

Chemical constituents:

Flavones, aglycones, triterpenoids, glycosides, carbohydrates, fatty acids, and volatile esters (Rao *et al.*, 2006). β-sitosterol, stigmasterol, flavones alkaloids, steroids, terpenoids, saponins, sugars, essential oil, resin, and tannin (Shree *et al.*, 2019).

Recipe:

Ingredients: Leaves, garlic, onion, gram flour, oil, chili powder, and salt.

Method: Wash and chop the leaves, then heat oil in a pan and fry them. Add crushed garlic cloves, followed by chopped vegetables, red chilies, and salt. Allow them to steam well. When the vegetables are halfway cooked, gradually sprinkle the dal flour and stir the vegetables continuously. Stirring evenly will help loosen the vegetables. Cook the vegetables on low flame.

Uses:

1. Skin conditions: The leaves and extracts of *C. halicacabum* are used in traditional medicine to alleviate skin conditions such as eczema, itching, and rashes. The plant possesses anti-inflammatory and antipruritic properties that may help soothe and relieve skin irritation (Sadique *et al.*, 1987).

2. Joint pain and inflammation: It is also used for its potential analgesic and anti-inflammatory properties. It is believed to help reduce joint pain, swelling, and inflammation associated with conditions like arthritis and rheumatism. The leaf of this plant mixed with castor oil is administered internally to treat rheumatism and

to check pain in the muscles and joints of the lower back (Kirthikar *et al.*, 1969).

3. Insect bites and stings: The crushed leaves are applied topically to soothe insect bites and stings. The plant's anti-inflammatory and antipruritic properties may help reduce itching, redness, and swelling caused by insect bites (Sadique *et al.*, 1987).

4. Traditional medicine: In traditional medicine systems, *C. halicacabum* has been used as a general tonic and to support overall health. It is believed to possess various medicinal properties, including anti-inflammatory, antiparasitic, anti-inflammatory, antipyretic, and diuretic effects (Sadique *et al.*, 1987; Asha *et al.*, 1999; Boonmars *et al.*, 2005).

5. Culinary: In some cultures, young shoots and leaves are consumed as a vegetable. They are often cooked and used in curries, stir-fries, or other dishes (Boonmars *et al.*, 2005).

CHAPTER 21

Wild Vegetables of the Family Smilacaceae

INTRODUCTION

The Smilacaceae family, or greenbrier family, includes around 320 species in the two main genera, *Smilax* and *Heterosmilax*. Mostly found in tropical and temperate regions, these climbing vines are recognized by their prickly stems and tendrils near the leaves. Though not widely used in cooking, some Smilax species have historical medicinal uses, notably sarsaparilla root (Conran 1998).

Smilax zeylanica L.

Botanical name: *Smilax zeylanica*

Family: Smilacaceae

Local name: Ghotvel

Vernacular name:

- **Assamese:** Kumarika
- **Bengali:** Hosti-karna lota
- **Malayalam:** Arikanni, Kareelanchi
- **Mishing:** Yorit

Season: Flower, Fruit: April-February

Parts used: Tuber, stem, root, and leaves

Characteristics (Kamble and Lobo, 2022):

1. Habit: *S. zeylanica* is a woody climber or vine that can grow up to several meters in length. It typically has a twining or trailing growth habit, using tendrils to attach itself to other plants or structures for support (Fig. **44.1**).

2. Leaves: The leaves of *S. zeylanica* are simple, alternate, and have a heart-shaped or ovate shape. They are usually dark green and glossy, with prominent veins. The leaf margins may be smooth or slightly toothed (Fig. **44.1**).

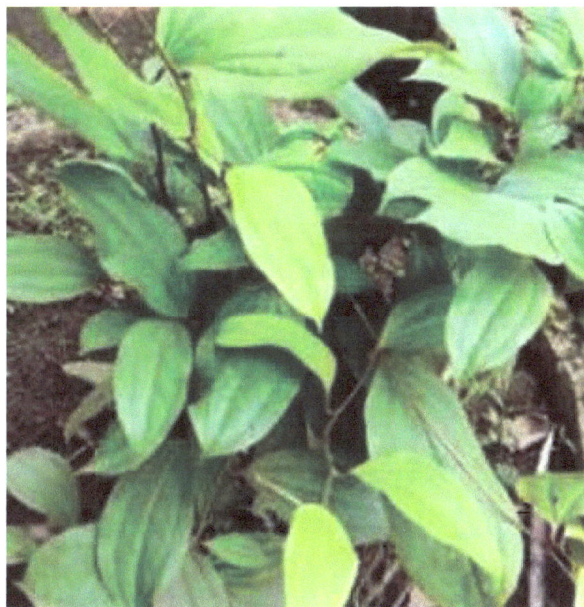

Fig. (44.1). Whole plants of *S. zeylanica* (PC: Madhura).

3. Stem: The stem of *S. zeylanica* is green and often has thorns or prickles along its length. It can be quite flexible and capable of winding around other plants or objects.

4. Flowers: *S. zeylanica* produces small, inconspicuous flowers that are typically greenish-yellow in color. The flowers are unisexual, meaning separate male and female flowers are borne on the same plant. They are arranged in clusters or umbels.

5. Fruits: After flowering, *S. zeylanica* develops small berries that turn from green to black when ripe. Each berry usually contains one or a few seeds.

Distribution:

India; Jawa; Malaya; Myanmar; Nepal; Solomon Is.; Sri Lanka

Propagation: Seeds.

Chemical constituents:

Steroidal saponins- smilasaponin, smilagenin, sarsasapogenin, and diosgenin, flavonoids- quercetin, kaempferol, and isorhamnetin, stilbenes- resveratrol, tannins, alkaloids, smilamine, smilacine, and smilacacidine (Madhavan *et al.*, 2010; Murali *et al.*, 2011; Rajesh *et al.*, 2014).

Recipe:

Ingredients: Younger stem, finely chopped onion, 5-6 garlic cloves, salt, green chilies, curry leaves, Asafoetida, mustard seeds, turmeric, gram dal, *etc.*

Method: Wash the curry leaves and chop them finely. Heat oil in a pan and fry asafoetida, mustard seeds, curry leaves, crushed garlic cloves, and finely chopped onions. Once the onions turn slightly yellow, add the chopped vegetables, chili, turmeric powder, soaked dal, and salt to taste. Lightly beat some water, cover the pan, and cook over low heat.

Uses:

1. Ayurvedic Medicine: In traditional Ayurvedic medicine, *S. zeylanica* has been used for its purported medicinal properties. It has shown diuretic, anti-inflammatory, and blood-purifying effects. It is sometimes used to support kidney health and as a remedy for skin disorders (Kumar *et al.*, 2005).

2. Culinary Uses: The tender shoots and young leaves of *S. zeylanica* are consumed as a vegetable in some cuisines. They are often cooked and used in stir-fries, soups, or curries. The plant has a slightly bitter and tangy taste (Kumar *et al.*, 2005).

3. Antioxidant and Anti-inflammatory Activity: *S. zeylanica* contains flavonoids and other compounds that have antioxidant and anti-inflammatory properties. These properties may contribute to its traditional use in promoting overall well-being (Sarvalingam *et al.*, 2016).

4. Traditional Remedies: In traditional practices, *S. zeylanica* has been used to alleviate conditions such as rheumatism, arthritis, and urinary tract infections (Jiang *et al.*, 2003; Jeyaprakash *et al.*, 2011; Shyma *et al.*, 2012; Breitbach *et al.*, 2013).

CHAPTER 22

Wild Vegetables of the Family Solanaceae

INTRODUCTION

The Solanaceae family, or nightshade family, includes over 2,000 species across 100 genera, ranging from herbs and shrubs to towering trees found worldwide. They are recognized by their star-shaped flowers with five petals and sepals, and alternate leaves. This family offers a mix of toxic and edible plants, including deadly nightshade, potatoes, tomatoes, and eggplants. While some have medicinal uses, caution is advised due to varying toxicity. Nutritionally, they provide vitamins and antioxidants, but some also contain harmful anti-nutrients (Chidambaram *et al.*, 2022).

Solanum nigrum L.

Botanical name: *Solanum nigrum*

Family: Solanaceae

Local name: Black Nightshade, Kamoni

Vernacular name:

- **Sanskrit:** Dhvansamaci
- **Bengali:** Gudakamai
- **English:** Garden night shade
- **Hindi:** Makoya, Kakamachi, Kali makoy
- **Kannada:** Ganikesopu
- **Malayalam:** Manatakkali
- **Marathi:** Kamoni
- **Punjabi:** Mako, Peelak, Mamoli
- **Urdu:** Mako

Season: December to March

Parts used: All parts

Ganesh Chandrakant Nikalje, Apurva Chonde, Sudhakar Srivastava & Penna Suprasanna

Characteristics:

1. Habit: *S. nigrum* is an annual or perennial herbaceous plant. It can grow upright or have a sprawling or prostrate growth habit, depending on the conditions (Dilip *et al.*, 2012).

2. Leaves: The leaves of *S. nigrum* are alternate and simple. They are typically ovate to lanceolate in shape, with serrated or wavy margins. The leaves are green and can have a smooth or slightly hairy surface (Dilip *et al.*, 2012).

3. Flowers: *S. nigrum* produces small, star-shaped flowers that are usually white, although they can also be pale yellow or purple. The flowers are borne in clusters or umbels and have five petals (Dilip *et al.*, 2012).

4. Fruits: After flowering, *S. nigrum* develops small berries that turn from green to black when ripe. The berries are round or slightly elongated and contain numerous small seeds (Dilip *et al.*, 2012) (Fig. **45.1**).

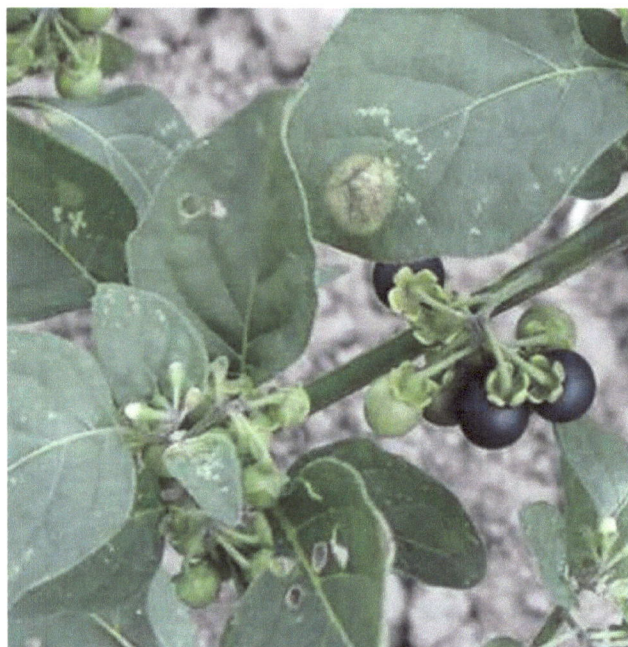

Fig. (45.1). Fruits of *S. nigrum* (PC: Apurva Shankar Chonde).

5. Stem: The stem of *S. nigrum* is usually green and can be slightly hairy or have sparse prickles. It can reach heights of up to one meter, depending on the growing conditions (Dilip *et al.*, 2012).

Distribution:

Europe, Asia- India, China, Japan, Thailand, Malaysia, Indonesia, and the Philippines, Africa- Nigeria, South Africa, Ethiopia, Kenya, Tanzania, and Egypt, Americas- North America, including the United States and Canada, as well as in Central and South America, Australia and Oceania- Australia, New Zealand, Fiji, and Papua New Guinea.

Propagation: Seed

Chemical constituents:

α and β-solamagrine, apigenin, caffeic acid, catechin, epicatechin, flavonoids, gallic acid, Gentisic acid, Kaempferol, Luteolin, m-coumaric acid, naringenin, protocatechuic acid, rutin, solanine, solamargine, solasonine, solasodine, solanidine, solasodinsolanidine, steroidal saponins and glycoprotein (Hoang *et al.*, 2014)

Recipe:

Ingredients: Kamoni, mustard seeds, cumin seeds, green chilies, salt, red chili powder, turmeric powder, coriander powder, coriander, *etc.*

Method: Heat oil in a pan on the stove. Keep the flame low. Once the oil is hot, add mustard seeds, cumin seeds, and green chilies. Fry them for a few seconds. Now add Kamoni and fry for 2 minutes. After frying the Kamoni, add salt, red chili powder, turmeric powder, and coriander powder. Fry the spices for 1 minute. Cover the pan and cook the tomatoes on medium heat, stirring occasionally. Once the Kamoni is cooked, add sugar and cook for an additional 2 minutes. Turn off the gas. Transfer the Kamoni sabji to a bowl. Garnish it with fresh coriander.

Uses:

1. Traditional Medicine: In traditional medicine systems, various parts of *S. nigrum*, including the leaves, berries, and roots, have been used for their potential medicinal properties. It is believed to possess anti-inflammatory, analgesic, diuretic, and antioxidant properties. It has been used to treat conditions such as fever, digestive disorders, skin problems, respiratory ailments, and urinary tract infections (Leporatti *et al.*, 2009; Jain *et al.*, 2011; Wang *et al.*, 2013).

2. Culinary: In certain cultures, *S. nigrum* is used as a culinary ingredient. The young leaves, shoots, and fruits are cooked and consumed as a vegetable in dishes such as stir-fries, curries, and soups. However, it is important to note that some varieties of *S. nigrum* may contain toxic compounds, particularly in unripe fruits

and leaves, so caution should be exercised when using it as food (Akubugwo *et al.*, 2007).

3. Ornamental Plant: Some varieties of *S. nigrum* are cultivated as ornamental plants for their attractive foliage and small berries. They can be grown in gardens or as potted plants to enhance the aesthetic appeal of outdoor spaces (Jain *et al.*, 2011).

4. Wildlife Food Source: The berries of *S. nigrum* serve as a food source for various birds and wildlife species. They are consumed by birds such as thrushes, blackbirds, and finches, contributing to biodiversity and ecological interactions (Jain *et al.*, 2011).

CHAPTER 23

Wild Vegetables of the Family Typhaceae

INTRODUCTION

The Typhaceae family, or cattail family, includes around 51 species in one or two genera (*Typha* and *Sparganium*). These wetland plants are recognized by their tall, emergent forms with long, strap-like leaves and dense flower spikes. Despite limited variety, Typhaceae plants have various uses. Some cultures use cattails in food preparations, and their dense pollen has been used as tinder. Medicinally, they have been used for wound healing, though research is limited. Nutritionally, they offer some carbohydrates, but their primary value lies in their ecological role in wetlands (Kubitzki 1998).

Typha latifolia L.

Botanical name: *Typha latifolia*

Family: Typhaceae

Local name: Cattail

- **Common name:** Broadleaf Reedmace, Broadleaf cattail, Bulrush, Common bulrush, Common cattail, Cat-o'-nine-tails, Great reedmace, Cooper's reed
- **Kashmiri:** Zab, Peit

Season: June- July

Parts used: All parts

Characteristics (Pojar *et al.*, 1994):

1. Habitat: *T. latifolia* is commonly found in wetland habitats, such as marshes, swamps, and the edges of lakes, ponds, and streams. It thrives in areas with freshwater or brackish water.

2. Growth Habit: *T. latifolia* is a perennial herbaceous plant that grows in dense clumps or colonies. It typically reaches a height of 3 to 9 feet (1 to 3 meters). The plant forms large stands of tall, erect stems.

3. Leaves: The leaves of *T. latifolia* are long and narrow, resembling strap-like blades. They are flat, smooth, and have parallel veins. The leaves arise from the base of the plant and grow in a tufted arrangement.

4. Flowers: The flowers of *T. latifolia* are arranged in dense cylindrical spikes, known as "catkins." The female catkins are located at the top of the spike and are brownish or dark in color. The male catkins are located below the female catkins and are yellowish in color. The female flowers are small and densely packed, while the male flowers are elongated and feathery (Fig. **46.1**).

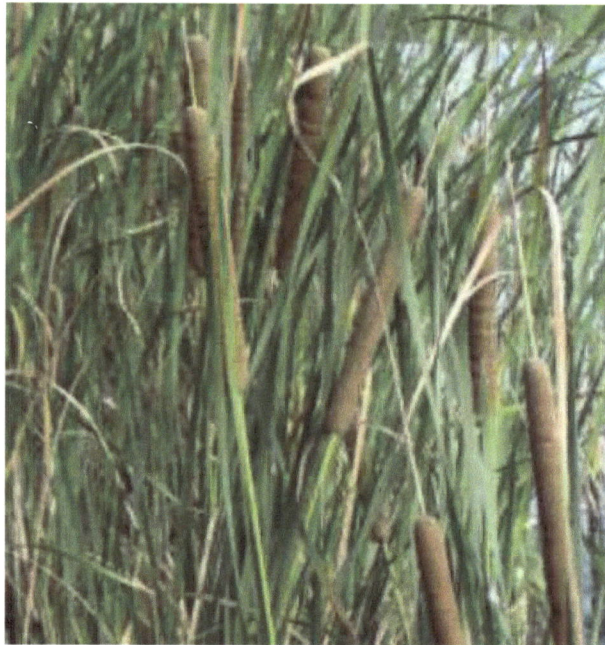

Fig. (46.1). Inflorescence of *T. latifolia* (PC: Apurva Shankar Chonde).

5. Inflorescence: The inflorescence of *T. latifolia* consists of separate male and female catkins. The female catkins are cylindrical and compact, while the male catkins are elongated and feathery.

6. Seeds: After pollination, the female flowers develop into small seeds enclosed in a fluffy, cotton-like material. These seeds are easily dispersed by wind or water.

Distribution:

North America- United States, Canada, and Mexico. Europe, Asia, China, Japan, India, Russia, Africa, and Australia.

Propagation: Seed, young shoots

Chemical constituents:

Phenolic Compounds- tannins, flavonoids, and phenolic acids, fatty acids- linoleic acid, palmitic acid, and oleic acid, proteins and amino acids, carbohydrates- starches and sugars, Minerals- potassium, calcium, magnesium, and trace elements such as iron and manganese.

Recipe:

Ingredients: Cattail hearts tender white bottom portions, white wine, vinegar, salt, bay leaf, lemon zest, oil, garlic, fresh thyme.

Method: Sauté the garlic in the oil, stirring occasionally, until it becomes aromatic and turns slightly tan in color. Be careful not to let the garlic burn. Add the shallot and continue cooking for 2 more minutes. Pour in the wine, cattail hearts, bay leaf, lemon, thyme, and salt. Cook the mixture covered for about 10 minutes, or until the cattails become tender. To prevent excessive evaporation and keep the pan moist, cover it with a piece of parchment paper. Finally, add the vinegar and taste the mixture for salt, adjusting as needed. Transfer the relish to a container with a tight-fitting lid and store it for future use. This relish can be kept for at least a month, if not longer.

Uses:

1. Food Source: The young shoots, tender stems, and rhizomes of *T. latifolia* can be consumed as a source of food. They are rich in starch and can be boiled, roasted, or ground into flour for making bread, porridge, or other food products (Pandey and Verma, 2018).

2. Medicinal Purposes: Various parts of *T. latifolia* have been used in traditional medicine. The plant has been employed for its diuretic, anti-inflammatory, and analgesic properties. It has been used to treat conditions such as urinary tract infections, dysuria, and kidney stones, and as a general detoxifying agent (Sesin *et al.*, 2021).

3. Craft and Construction Material: The leaves and stems of *T. latifolia* have been traditionally used for making handicrafts, such as baskets, mats, and hats. The

strong and flexible nature of the plant's fibers makes them suitable for weaving and construction purposes (Sesin *et al.*, 2021).

4. Water Filtration: Cattails, including *T. latifolia*, are known for their ability to absorb pollutants and purify water. They can be used in constructed wetlands or natural water bodies to help improve water quality by filtering out contaminants (Pandey and Verma, 2018).

5. Wildlife Habitat: *T. latifolia* provides important habitat and food sources for various wildlife species, including birds, insects, amphibians, and small mammals. The dense stands of cattails offer nesting sites, shelter, and food for these organisms (Rook 2004).

6. Soil Stabilization: The extensive root system of *T. latifolia* helps stabilize soil and prevent erosion. It can be utilized in erosion control projects, especially in wetland restoration and shoreline stabilization efforts (Sesin *et al.*, 2021).

CHAPTER 24

Wild Vegetables of the Family Verbinaceae

INTRODUCTION

The Verbenaceae family, or Verbena family, includes over 1200 species in around 34 genera. Primarily tropical, these aromatic plants range from herbs and shrubs to trees. They feature opposite or whorled leaves and small flowers with four or five petals in spikes, clusters, or racemes. Popular for ornamental and historical medicinal uses, the family is best known for lemon verbena, which adds a citrusy touch to dishes (Atkins 2004).

Clerodendrum serratum (L.) Moon.

Botanical name: *Clerodendrum serratum*

Family: Verbenaceae

Local name: Bharangi

Vernacular name:

- **Hindi:** Bharangi
- **Sanskrit:** Bharangi
- **English:** Bharangi

Season: August to October

Parts used: Roots and Leaves

Characteristics:

1. Habit: *C. serratum* is a perennial shrub or vine that can grow up to 2-3 meters in height. It has a woody stem with branches.

2. Leaves: The leaves of *C. serratum* are opposite, simple, and elliptic to lanceolate in shape. They have serrated margins and are approximately 5-15 cm in length. The leaves are dark green and glossy in appearance.

Ganesh Chandrakant Nikalje, Apurva Chonde, Sudhakar Srivastava & Penna Suprasanna

3. Flowers: The flowers of *C. serratum* are tubular and arranged in terminal clusters called cymes. They have a distinct blue or violet color with a white throat. The corolla tube is about 2 cm long and curved, while the four petals are spreading (Fig. **47.1**).

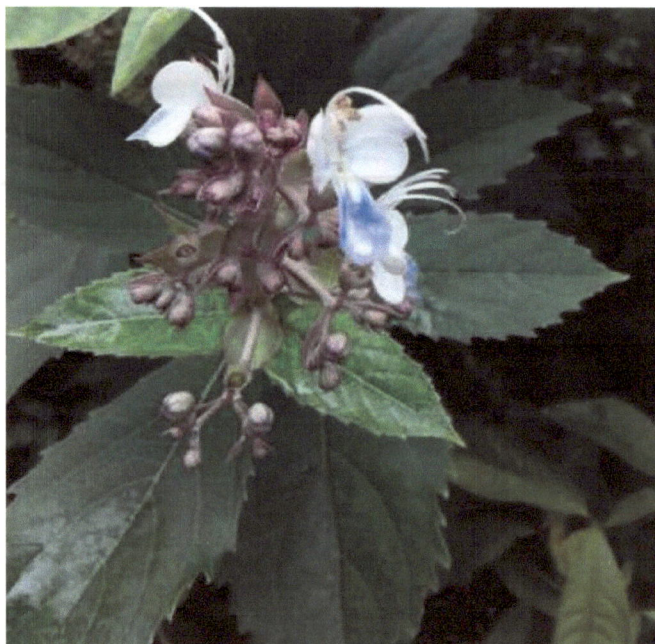

Fig. (47.1). Inflorescence of *C. serratum* (PC: Apurva Shankar Chonde).

4. Fruits: After flowering, *C. serratum* produces small, rounded fruits that are about 6-8 mm in diameter. The fruits are initially green and turn black when mature.

Distribution:

The plant is distributed over scrub forests throughout the tropical and sub-tropical parts up to 1500 m, particularly in Bengal, Odisha and peninsular India.

Propagation: Stem cuttings of semi-hardwood.

Chemical constituents:

Some of the major constituents present in the plant include D-mannitol, hispidulin, cleroflavone, apigenin, scutellarein, serratagenic acid, acteoside, verbascoside, oleanolic acid, clerodermic acid, γ-sitosterol, β-sitosterol, cholestanol, clerosterol, campesterol, and 24-ethyl cholesterol. Additionally, it contains compounds like 7-β-coumaroyl-oxyugandoside and 7-βcinnamoy--oxyugandoside.

Recipe:

Ingredients: Younger leaves, finely chopped onion, 5-6 garlic cloves, salt, green chilies, curry leaves, Asafoetida, grated coconut, mustard seeds, turmeric, Jaggery, gram dal, *etc.*

Method: First, peel the stalks of the leaves and wash a single Judy leaves. Chop the leaves finely. Heat oil in a pan and fry asafoetida, mustard seeds, curry leaves, crushed garlic cloves, and finely chopped onion. Once the onion turns slightly yellow, add the chopped vegetables, chili, turmeric powder, soaked dal, jaggery, and salt to taste. Lightly beat water and cover the pan. Cook the mixture over low heat until the vegetables are tender. Once cooked, garnish with freshly grated wet coconut.

Uses:

1. Respiratory Disorders: In traditional medicine, *C. serratum* is used for the treatment of respiratory disorders such as cough, asthma, and bronchitis. It is believed to have expectorant and bronchodilatory properties (Jagruti *et al.*, 2014).

2. Anti-inflammatory and Analgesic Activity: *C. serratum* has shown potential anti-inflammatory and analgesic effects in experimental studies. It is used to alleviate pain and inflammation associated with various conditions (Narayanan *et al.*, 1999).

3. Anti-diabetic Properties: The leaves and roots of *C. serratum* have been traditionally used to manage diabetes. It is believed to have hypoglycemic activity and may help regulate blood sugar levels (Chae *et al.*, 2006).

4. *C. serratum* is utilized in the treatment of serious diseases such as syphilis, typhoid, cancer, jaundice, hypertension, asthma, and inflammatory and infectious disorders (Jagruti *et al.*, 2014).

5. The isolated flavonoids, such as hispidulin and cleroflavone, demonstrate significant activities including antioxidant, antimicrobial, anti-asthmatic, anti-tumor, and CNS-binding properties (Shi *et al.*, 1993; Rajlakshmi *et al.*, 2003, Chae *et al.*, 2006; Vincent *et al.*, 2012).

6. The root of *C. serratum* holds a significant value as one of the key components of Brahata panchamool, a traditional formulation, and is in high demand (Chae *et al.*, 2006).

CHAPTER 25

Wild Vegetables of the Family Vitaceae

INTRODUCTION

The Vitaceae family, or grape family, includes over 750 species across 14 genera, primarily in tropical and warm temperate regions. These climbing plants are identified by swollen leaf nodes and tendrils. Tiny flowers clustered opposite the leaves yield the family's most famous product grapes. Enjoyed fresh, dried, or juiced, grapes are a global culinary staple. While some Vitaceae species have medicinal uses, caution is advised as some contain toxins. Nutritionally, grapes are rich in vitamins and antioxidants, while other family members have limited nutritional value (Wen 2007).

Leea indica (Burm.f.) Merr.

Botanical name: *Leea indica*

Family: Vitaceae

Local name: Dinda

Vernacular name:

- **Common name:** Bandicoot Berry
- **Hindi:** Kukur jihwa
- **Manipuri:** Koknal
- **Marathi:** Karkani
- **Tamil:** Nalava, Ottannalam
- **Malayalam:** Nakku
- **Telugu:** Amkador
- **Kannada:** Andilu, Tannunuka, Gadapatri
- **Bengali:** Kurkur
- **Assamese:** Ahina
- **Sanskrit:** Chatri
- **Mizo:** Kawlkar

Season: Flowering and fruiting: March-August

Parts used: Leaves, roots, stem bark, inflorescence, and flowers.

Characteristics (Balkrishna *et al.*, 2023):

1. Growth Habit: *L. indica* is a medium-sized shrub that can grow up to 2-3 meters in height. It has a spreading and bushy growth habit with multiple branches.

2. Leaves: The leaves of *L. indica* are compound, alternate, and large in size. Each leaf is composed of 5-7 leaflets arranged in a palmate pattern. The leaflets are ovate or lanceolate in shape, with serrated or toothed margins. The upper surface of the leaves is usually dark green, while the lower surface may have a lighter shade (Fig. **48.1**).

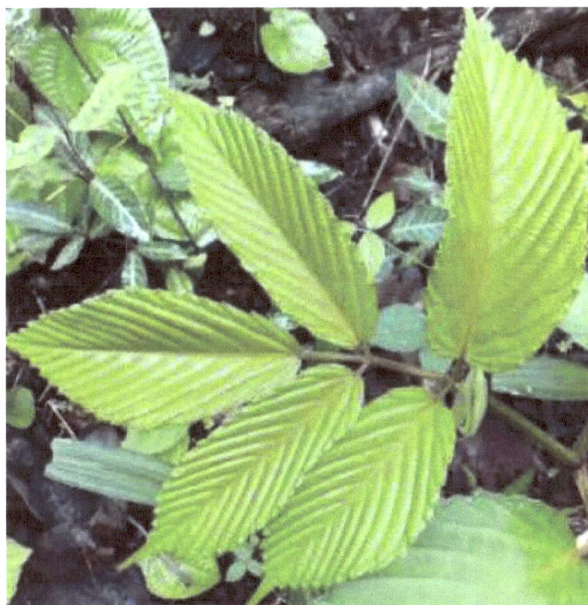

Fig. (48.1). Leaves of *L. indica* (PC: Apurva Shankar Chonde).

3. Flowers: *L. indica* produces small, inconspicuous flowers that are arranged in panicles or clusters at the ends of the branches. The flowers are typically greenish-white to pale yellow in color.

4. Fruits: The fruits of *L. indica* are small berries that turn from green to dark purple or black when ripe. The berries are round or slightly oval in shape and may contain several seeds.

5. Stem and Bark: The stems of *L. indica* are slender and woody, with a brownish or grayish bark. The bark may have a rough texture.

Distribution:

Southeast Asia- Thailand, Myanmar, Laos, Cambodia, Vietnam, Malaysia, Indonesia, and the Philippines, Indian Subcontinent- India, Nepal, Bangladesh, and Sri Lanka, China- Yunnan and Guangxi regions, *etc.*

Propagation: Seed, cuttings, air layering

Chemical constituents:

Alkaloids, flavonoids, gallic acid, glycosides, lupeol, mollic acid, arabinoside, mollic acid, xyloside, quercetin, saponins, steroids, terpenoids, ursolic acid, β-sitosterol (Kekuda *et al.*, 2018).

Recipe:

Ingredients: Younger leaves, finely chopped onion, 5-6 garlic cloves, salt, green chilies, curry leaves, Asafoetida, grated coconut, mustard seeds, turmeric, besan, *etc.*

Method: Wash the leaves and chop them finely. Heat oil in a pan and fry asafoetida, mustard seeds, curry leaves, crushed garlic cloves, and finely chopped onion. When the onion turns slightly yellow, add the chopped vegetables, chili, and turmeric powder. Also, add besan (gram flour) and salt to taste. Lightly beat water and cover the pan. Cook over low flame. Once the vegetables are cooked, add wet coconut on top.

Uses:

1. Medicinal Purposes: In traditional medicine practices, different parts of *L. indica*, such as the leaves, roots, and bark, are used to treat various ailments. It is believed to have properties such as anti-inflammatory, analgesic, antipyretic, and antioxidant effects. It is used for conditions like fever, cough, cold, skin diseases, rheumatism, and digestive disorders (Burkill, 1966; Rahman *et al.*, 2013; Siew *et al.*, 2014; Singh *et al.*, 2019).

2. Culinary: In some regions, the leaves of *L. indica* are used as a culinary ingredient. They are added to dishes like soups, stews, and curries to enhance flavor and provide nutritional value (Rahman *et al.*, 2013).

3. Traditional Health Tonic: *L. indica* is sometimes prepared as a traditional health tonic or herbal tea. It is believed to have general health-promoting properties and is consumed for its potential benefits on overall well-being (Singh *et al.*, 2019).

4. Ornamental Plant: *L. indica* is also grown as an ornamental plant in gardens and landscapes. Its attractive foliage and colorful berries make it an appealing addition to gardens (Burkill, 1966).

Leea macrophylla Roxb. ex Hornem.

Botanical name: *Leea macrophylla*

Family: Vitaceae

Local name: Dinda

Vernacular name:

- **Assamese:** Kath tenga
- **English:** Hathikana
- **Malayalam:** Njallu
- **Other:** Hathikana

Season: Flowering: August – October; Fruit ripe: October - March

Parts used: Leaves, Fruits.

Characteristics (Mahato and Sharma 2019):

1. Growth Habit: *L. macrophylla* is a large shrub or small tree that can reach heights of up to 3-6 meters. It has a dense and bushy growth habit with multiple branches.

2. Leaves: The leaves of *L. macrophylla* are large, hence the species name "macrophylla," which means large leaves. The leaves are compound and palmately lobed, consisting of three to five leaflets. The leaflets are ovate or elliptic in shape, with serrated or toothed margins. The upper surface of the leaves is typically dark green, while the lower surface may have a lighter shade.

3. Inflorescence: *L. macrophylla* produces small flowers that are arranged in large, terminal panicles. The flowers are typically greenish-yellow or pale yellow in color (Fig. **48.2**).

Fig. (48.2). Inflorescence of *L. macrophylla* (PC: Apurva Shankar Chonde).

4. Fruits: The fruits of *L. macrophylla* are small berries that turn from green to red or black when ripe. The berries are round or slightly oval in shape and may contain several seeds.

5. Stem and Bark: The stems of *L. macrophylla* are slender and woody, with a grayish or brownish bark. The bark may have a slightly rough texture.

Distribution:

E. Asia - southwest China, India, Myanmar, Thailand, Cambodia, Laos, Vietnam.

Propagation: Seed, Cuttings, Air layering

Chemical constituents: *L. macrophylla* is reported to be rich in various phytochemical constituents. According to a study conducted by Malik and Upadhyay (2020), the plant contains alkaloids, steroids, glycosides, saponins, carbohydrates, proteins, and tannins. In addition, the leaves of *L. macrophylla* are known to contain a significant amount of phenolic constituents. These phenolic compounds include flavonoids, leucoanthocyanidins, p-hydroxybenzoic acid, syringic acid, and gallic acid.

Recipe:

Ingredients: Young leaves, finely chopped onion, 5-6 garlic cloves, salt, green chilies, curry leaves, Asafoetida, mustard seeds, turmeric, and chilies.

Method: Wash a handful of fresh young leaves. Heat oil in a pan and fry asafoetida, mustard seeds, curry leaves, crushed garlic cloves, and finely chopped onions and chilies. When the onions turn slightly yellow, add the chopped vegetables, along with chili powder, turmeric powder, and salt to taste. Lightly beat some water, cover the pan, and cook over low heat.

Uses:

1. Medicinal Purposes: In traditional medicine, various parts of *L. macrophylla*, such as leaves, roots, and stems, are used to treat different ailments. It is believed to possess anti-inflammatory, analgesic, and antipyretic properties. Extracts from the plant have been used to alleviate pain, reduce fever, gastric tumor, goiter, lipoma, and tetanus, treat skin disorders, and promote wound healing (Zaoui *et al.*, 2002; Garodia *et al.*, 2007).

2. Culinary: The young leaves of *L. macrophylla* can be used as a food source. They are sometimes consumed as a leafy green vegetable and can be added to soups, stews, or stir-fries (Jadhao *et al.*, 2008; Banik *et al.*, 2014; Raghunathan, 2017).

3. Ornamental Plant: With its attractive foliage and vibrant red or purple berries, *L. macrophylla* is also cultivated as an ornamental plant in gardens and landscapes. Its large, glossy leaves and colorful berries add aesthetic value to outdoor spaces.

4. Traditional Dye: The leaves of *L. macrophylla* can be used to create natural dyes. They contain pigments that can be extracted and used to color fabrics, yarns, and other materials (Banik *et al.*, 2014).

5. Environmental Uses: *L. macrophylla* has potential environmental applications. It is known to have allelopathic properties, meaning it can release certain chemicals that inhibit the growth of other plants nearby. This characteristic has led to its use in weed management and controlling invasive plant species (Banik *et al.*, 2014).

CHAPTER 26

Wild Vegetables of the Family Zingiberaceae

INTRODUCTION

The Zingiberaceae family, or ginger family, includes over 1,300 species across 50 genera, primarily in tropical and subtropical regions. These aromatic perennial herbs are recognized by their underground rhizomes, which are often the most valuable part. Key features include the absence of latex and the presence of stipules at the base of leaves. Culinary stars like ginger and turmeric dominate, while cardamom and galangal add unique flavors. Medicinally, ginger aids digestion, and turmeric has anti-inflammatory properties. Nutritionally, the family offers vitamins, minerals, and bioactive compounds, though some members may contain anti-nutrients (Larsen *et al.*, 1998).

Curcuma neilgherrensis Wight

Botanical name: *Curcuma neilgherrensis*

Family: Zingiberaceae

Local name: Ranhalad

Vernacular name:

- **Common name:** Nilgiri Turmeric
- **Malayalam:** Koova, Kattumanjal, Vellakkuva
- **Marathi:** Ranhalad

Season: June to October

Parts used: All parts

Characteristics:

1. Plant Height: *C. neilgherrensis* is a relatively tall plant, with mature plants reaching heights of around 1 to 2 feet (30 to 60 centimeters). However, the exact height can vary depending on growing conditions (Shyam *et al.*, 2012).

2. Rhizomes: *C. neilgherrensis* has thick, fleshy rhizomes that grow underground. The rhizomes are aromatic and contain the bioactive compounds responsible for their medicinal properties (Shyam *et al.*, 2012).

3. Leaves: The plant has long, lance-shaped leaves that arise directly from the ground. The leaves are green, glossy, and have prominent veins. They grow in an alternate arrangement along the stem (Shyam *et al.*, 2012).

4. Inflorescence: *C. neilgherrensis* produces an inflorescence that emerges from the base of the stem. The inflorescence consists of a central spike surrounded by colorful bracts. The bracts can be various shades of pink, purple, or white, and they form an attractive cluster (Shyam *et al.*, 2012).

5. Flowers: The individual flowers of *C. neilgherrensis* are small and tubular. They are typically white or pale yellow in color. The flowers are arranged in dense, cone-like structures on the central spike of the inflorescence (Shyam *et al.*, 2012) (Fig. **49.1**).

Fig. (**49.1**). **Whole plant of** *C. neilgherrensis* (PC: Apurva Shankar Chonde).

Distribution:

It is native to the Nilgiri Hills of Southern India. It is primarily found in the states of Tamil Nadu and Kerala.

Propagation: Rhizome

Chemical constituents:

Curcuminoids, essential oils, turmerones, flavonoids, quercetin and kaempferol, phenolic acids and their derivatives, and caffeic acid derivatives.

Recipe:

Ingredients: Young tender leaves, onion, garlic, gram masala, ground onion-coconut paste, oil, salt, masoor, black peas, *etc.*

Method: Heat oil in a pan and add the onions. Fry until they turn a golden-brown color. Add salt, masoor dal or black peas, and garam masala. Once the masoor dal or black peas are cooked, add the onion-coconut paste, water, and tender leaves. Cook the mixture until the leaves become tender.

Uses:

1. Culinary: The rhizomes of *C. neilgherrensis* are used as a culinary spice, similar to other turmeric varieties. It imparts a distinct flavor and vibrant yellow color to various dishes (Chaithra *et al.*, 2013).

2. Medicinal: *C. neilgherrensis* possesses medicinal properties and is used in traditional medicine systems. It is believed to have anti-inflammatory, antioxidant, and antimicrobial properties. It is used to alleviate digestive issues, promote wound healing, chronic hepatitis, anti-arthritis, antiseptic, and menstrual disorders, and support overall health (Gantait *et al.*, 2011).

3. Cosmetics and Skincare: The extract of *C. neilgherrensis* is used in cosmetics and skin care products due to its skin-brightening and anti-aging properties. It is used in creams, lotions, and face masks to improve skin tone and texture (Chaithra *et al.*, 2013).

4. Natural Dye: The vibrant yellow color of *C. neilgherrensis* makes it suitable as a natural dye. It is used to dye fabrics, yarns, and other materials (Chaithra *et al.*, 2013).

5. Ornamental Plant: *C. neilgherrensis* is also cultivated as an ornamental plant for its attractive foliage and flowers. It adds beauty to gardens and landscapes (Chaithra *et al.*, 2013).

CHAPTER 27

Wild Vegetables of the Family Zygophyllaceae

INTRODUCTION

The Zygophyllaceae family, or caltrop family, includes around 240 species in 22 genera, thriving primarily in dry and hot climates, including deserts and saline environments. These plants are characterized by shrubby or herbaceous forms with opposite or spirally arranged leaves, sometimes spiny or fleshy. Their flowers feature four or five separate petals and sepals, typically with eight to ten stamens. While some species have medicinal uses, culinary contributions are limited. Some members contain nutritional compounds such as steroidal saponins, flavonoids, alkaloids, terpenoids, etc. (Sheahan 2007).

Tribulus terrestris L.

Botanical name: *Tribulus terrestris*

Family: Zygophyllaceae

Local name: Gokharu

Vernacular name:

- **Sanskrit:** Gokshur
- **English:** Caltrops
- **Hindi:** Gokharu
- **Gujarathi:** Bethagokharu or Nanagokharu
- **Tamil:** Nerinjil
- **Urdu:** Khar-e-khusak khurd

Season: January to April

Parts used: All parts

Characteristics (Chhatre *et al.*, 2014):

Ganesh Chandrakant Nikalje, Apurva Chonde, Sudhakar Srivastava & Penna Suprasanna

1. Growth Habit: *T. terrestris* is a low-growing, spreading herbaceous plant. It has a prostrate or decumbent growth habit, meaning it lies flat on the ground or spreads along the surface.

2. Leaves: The leaves of *T. terrestris* are small, opposite, and pinnately compound. Each leaflet is typically elliptical or lanceolate in shape and has a spiny margin. The leaves are arranged in pairs along the stems.

3. Stems: The stems of *T. terrestris* are slender, wiry, and covered with short, stiff hairs. They often branch near the base and can trail along the ground.

4. Flowers: *T. terrestris* produces solitary flowers that are small and typically yellow in color. The flowers have five petals and are radially symmetrical. They have a star-like appearance and are borne on short stalks arising from the leaf axils (Fig. **50.1**).

Fig. (50.1). Flowering twig of *T. terrestris* (PC: Trilok Barge).

5. Fruits: After flowering, *T. terrestris* develops distinctive spiny fruits known as burrs or nutlets. These fruits consist of several hard, woody segments with sharp spines. The spines are adapted for attachment and dispersal, as they can easily stick to clothing or animal fur.

Distribution:

Asia- India, Pakistan, Sri Lanka, Bangladesh, Nepal, Myanmar (Burma), Thailand, Malaysia, Indonesia, China, Japan, and the Middle East. Europe- Greece, Bulgaria, Romania, and Ukraine. Africa- Egypt, Sudan, Ethiopia, Kenya, Tanzania, Uganda, Nigeria, South Africa, and Madagascar. North America- United States, including Arizona, California, and Nevada, Australia, South America- Argentina, Brazil, and Chile.

Propagation: Seeds

Chemical constituents:

Steroidal saponins- protodioscin and dioscin, flavonoids- kaempferol, quercetin, apigenin, and rutin, alkaloids- harmine, harmaline, and harmane, terpenoids- saponins and triterpenes, phenolic compounds- gallic acid and ellagic acid, sterols- beta-sitosterol and stigma-sterol.

Recipe:

Ingredients: Tender leaves, soaked moogdal or turdal, finely chopped onion, cumin, mustard, Asafoetida, chili powder, salt, *etc.*

Recipe: Wash and chop the leaves. Heat oil in a pan and add cumin seeds, mustard seeds, and asafoetida. Fry the lentils (dal) and onions until golden brown. Then, add the chopped vegetables and stir well. Sprinkle chili powder and salt to taste. Next, add some water and cook the vegetables until tender.

Uses:

1. Enhancing Athletic Performance: *T. terrestris* is often used as a dietary supplement by athletes and fitness enthusiasts. It is believed to have potential benefits for increasing strength, stamina, and muscle mass. Some individuals use it to support athletic performance and improve exercise capacity (Kalamegam *et al.*, 2008).

2. Sexual Health and Libido: *T. terrestris* is widely known for its traditional use as an aphrodisiac. It has been used to enhance sexual desire, improve erectile function, and increase overall sexual performance in both men and women. It is a common ingredient in many herbal supplements targeted towards sexual health (Bitzer *et al.*, 2013).

3. Hormonal Balance: *T. terrestris* is believed to have an influence on hormone levels, particularly testosterone. It is sometimes used to support hormonal balance and regulate the production of reproductive hormones (Kalamegam *et al.*, 2008).

4. Kidney and Urinary Health: In traditional medicine, *T. terrestris* has been used to support kidney health and promote urine production. It may have diuretic properties, which may help in flushing out toxins and preventing urinary tract infections (Kalamegam *et al.*, 2008).

5. Antioxidant and Anti-inflammatory Effects: *T. terrestris* contains various bioactive compounds with antioxidant and anti-inflammatory properties. These properties may help protect the body against oxidative stress and inflammation, which are linked to various health conditions (Kalamegam *et al.*, 2008).

CONCLUSION

Biodiversity is as important to the life on Earth as is the oxygen and water. This fact must be realized and spread to every corner of the society and to all the people. The present book explores the diversity of wild vegetables in India and presents a repertoire of resourceful information on "124 wild vegetables" belonging to 50 different families. The diversity of vegetables on our plate is diminishing with the 'same' vegetable being repeated on a regular basis. The book presents information lucidly on alternate potential vegetables that are available in nature. With a diverse group of vegetables on our plate, greater diversity in tastier food and enriched nutrition can be possible. Most importantly, the diversity brings in nutritional requirements more effectively as compared to the consumption of a few specific vegetables.

Nutritional deficiency, for example of minerals iron (Fe) and zinc (Zn) and vitamins like vitamin A, are common throughout the world. It is partly due to the inadequate quantity of food consumed and partly due to the consumption of very less variety of vegetables and, some specific common vegetables and fruits. The present book demonstrates the rich diversity of several vegetables, the "wild, edible plants" that can be easily adopted and consumed by the people. With the increase in modernization and globalization, there is a rapid disappearance of traditional things including local food and traditional delicacies. There are recipes known for the wild vegetables and this book has attempted to provide information on this culinary angle. The readers of the book shall become aware of the vast diversity of vegetables they can consume and can also gain benefit from known recipes. Obviously, there may be many more ways to consume such vegetables, such as soup, salad, or other delicacies.

Another important point to consider is the increasing ailments and mental and physical disorders in the present era of environmental pollution, contaminated food items, and unhealthy lifestyle. It is worth noting that wild vegetables also possess several medicinal properties and plenty of natural phytochemicals. It is a well-known fact that in earlier times, people used a number of plant-based traditional medicines throughout the world, and that knowledge is vastly respected even today. In fact, the knowledge of traditional medicine has paved the way for several drugs used in the present era. The presented data on wild vegetables also includes their usefulness in terms of medicinal value. The known uses of various wild vegetables for medicinal purposes are included in the book.

In conclusion, the present book offers the readers a valuable resource on the wild vegetables' diversity of India along with their various uses. The book chapters include scientific to traditional information ranging from classification, chemical composition, and cooking instructions to medicinal applications.

REFERENCES

Abdel Kawy, MA (2023). An overview on botanical characteristics, phytochemical constituents and pharmacological effects of *Cordia dichotoma* (G. Forst.) and *Cordia sebestena* (L.) (Boraginaceae). *Egyptian Journal of Chemistry.*

Acevedo-Rodríguez, P., Strong, M.T. (2012). Catalogue of seed plants of the West Indies. *Smithson. Contrib. Bot., 98,* 1-1192.
[http://dx.doi.org/10.5479/si.0081024X.98.1]

Acharya, Y.T. (2011). *Agnivesha. Commentary by Chakrapanidatta. Charakasamhita of Charaka, Chikitsasthana. 14, Ver.* (Vol. 133-135, p. 507). Varanasi: Chaukhamba Press.

Acharya, Y.T. (2011). *Agnivesha. Commentary by Chakrapanidatta. Charakasamhita of Charaka, Chikitsasthana. 19, Ver.* (Vol. 31-33, p. 551). Varanasi: Chaukhamba Press.

Adebanjo, A.O., Adewama, C.O., Essein, E.E. (1983). Anti- infective agents of higher plants. Proceedings of an international symposium of Medicinal plants, Fifth edition, University of Ife *Nigeria,* 152-158.

Agrawal, D.P. (2002). Complementary and alternative medicine: An overview. *Curr. Sci., 82,* 518-524.

Aguiar, S., Borowski, T. (2013). Neuropharmacological review of the nootropic herb *Bacopa monnieri. Rejuvenation Res., 16*(4), 313-326.
[http://dx.doi.org/10.1089/rej.2013.1431] [PMID: 23772955]

Ahirrao, RA, Borsel, LB, Desai, SG (2009). Evaluation of anti-inflammatory activity of Corchorus trilocularis linn. *Seed oil, Indian Science abstracts, 17512*(45), 18.

Ahmed, A.S., Elgorashi, E.E., Moodley, N., McGaw, L.J., Naidoo, V., Eloff, J.N. (2012). The antimicrobial, antioxidative, anti-inflammatory activity and cytotoxicity of different fractions of four South African Bauhinia species used traditionally to treat diarrhoea. *J. Ethnopharmacol., 143*(3), 826-839.
[http://dx.doi.org/10.1016/j.jep.2012.08.004] [PMID: 22917809]

Ahmed, F., Urooj, A. (2010). Traditional uses, medicinal properties, and phytopharmacology of *Ficus racemosa* : A review. *Pharm. Biol., 48*(6), 672-681.
[http://dx.doi.org/10.3109/13880200903241861] [PMID: 20645741]

Akacha, L., Dikko, J., Khan, M., Anyam, J., Igoli, J. (2016). Phytochemical screening and antimicrobial activity of *Bryophyllum pinnatum* extracts. *Br. Biotechnol. J., 16*(2), 1-8.
[http://dx.doi.org/10.9734/BBJ/2016/28905]

Akpapunam, M.A., Sefa-Dedeh, S. (1997). Jack bean (*Canavalia ensiformis*): Nutrition related aspects and needed nutrition research. *Plant Foods Hum. Nutr., 50*(2), 93-99.
[http://dx.doi.org/10.1007/BF02436029] [PMID: 9201744]

Akubugwo, I.E., Obasi, A.N., Ginika, S.C. (2007). Nutritional potential of the leaves and seeds of black nightshade- *Solanum nigrum* L. Var virginicum from Afikpo-Nigeria. *Pak. J. Nutr., 6*(4), 323-326.
[http://dx.doi.org/10.3923/pjn.2007.323.326]

Alam, F., Ghosh, S., Pathak, B.J., Judder, M.I., Zaman, H., Sarma, A., Sarkar, D., Yakin, J. (2022). Nutraceuticals and Anti-Inflammatory Properties of *Basella Rubra*: A Review. *J. Pharm. Negat. Results, 28,* 5835-5839.

[http://dx.doi.org/10.47750/pnr.2022.13.S07.708]

Aleem, M. (2020). Anti-inflammatory and antimicrobial potential of *Plumbago zeylanica* L: A review. *J. Drug Deliv. Ther., 10*(5-s), 229-235.
[http://dx.doi.org/10.22270/jddt.v10i5-s.4445]

Al-Fatimi, M., Wurster, M., Schröder, G., Lindequist, U. (2007). Antioxidant, antimicrobial and cytotoxic activities of selected medicinal plants from Yemen. *J. Ethnopharmacol., 111*(3), 657-666.
[http://dx.doi.org/10.1016/j.jep.2007.01.018] [PMID: 17306942]

Al-Snafi, A.E (2019). A review on *Luffa acutangula*: A potential medicinal plant. *OSR J. Pharm., 9*(9), 56-67.

Almaaty, A.H., Keshk, S., Galal, A., Abbas, O.A., Hassan, M.K. (2022). Medicinal usage of some Arecaceae family members with potential anticancer effect. *J. Biotech Res., 13*, 55-63.

Almeida, M. (1993). *Flora of Maharashtra, Blatter Herbarium..* Mumbai: St. Xavier's College.

Al-Snafi, A.E. (2015). The pharmacological importance of *Asparagus officinalis*- A review. *J. Pharma. Biol., 5*(2), 93-98.

Al-Yousef, H.M., Hassan, W.H.B., Abdelaziz, S., Amina, M., Adel, R., El-Sayed, M.A. (2020). UPLC-ES--MS/MS Profile and Antioxidant, Cytotoxic, Antidiabetic, and Antiobesity Activities of the Aqueous Extracts of Three Different *Hibiscus* Species. *J. Chem., 2020*, 1-17.
[http://dx.doi.org/10.1155/2020/6749176]

Amat, N., Amat, R., Abdureyim, S., Hoxur, P., Osman, Z., Mamut, D., Kijjoa, A. (2014). Aqueous extract of dioscorea opposita thunb. normalizes the hypertension in 2K1C hypertensive rats. *BMC Complement. Altern. Med., 14*(1), 36.
[http://dx.doi.org/10.1186/1472-6882-14-36] [PMID: 24447776]

Amit, L., Kumar, C.A., Vikas, G., Parveen, B., Renu, B. (2010). Phytochemistry and pharmacological activities of *Capparis zeylanica*: an overview. *Int. J. Res. Ayurveda Pharm., 1*(2), 384-389.

Amritveer, B., Junaid, N., Bajwa, N. (2016). A review update on *Dillenia indica*. F. Elongata (MIQ.). *J. Drug Deliv. Ther., 6*(2), 62-70.
[http://dx.doi.org/10.22270/jddt.v6i2.1226]

Anagha, A.R., Anuradha, S.U., Taware, S.P. (2016). Bioactivity of indigenous plant *Glossocardia bosvallia* (L.f.) DC. against insect pests of stored products. *Ind J Tradit. Knowl., 15*(2), 260-265.

Anand, M., Basavaraju, R. (2021). A review on phytochemistry and pharmacological uses of *Tecoma stans* (L.) Juss. ex Kunth. *J. Ethnopharmacol., 265*, 113270.
[http://dx.doi.org/10.1016/j.jep.2020.113270] [PMID: 32822823]

Anjana, M., Swathi, V., Ramya, S.A., Divya, N., Sunisha, Y.A. (2020). Review on *Momordica dioica* fruits. *J Adv Plant Sci., 3*(1), 1-5.

Anonymous, (2007). The wealth of India: first supplement series (raw materials). *National Institute of Science Communication and Information Resources, CSIR, 2*, 211-213.

Anonymus. (1982), Tamarind juice concentrate plant starts in Mysore. Indian Food Industry. Mysore, India.

Arackal, E.J., Pandurangan, A.G. (2015). Taxonomy, conservation and sustainable utilisation of the genus *Dioscorea* in western. *J. Root Crops, 41*(2), 3-16.

Arinathan, V., Mohan, V.R., Britto, A., Murugan, C. (2007). Wild edibles used by Palliyars of the western Ghats, Tamil Nadu. *Indian J. Tradit. Knowl., 6,* 163-168.

Arinathan, V., Mohan, V.R., Maruthupandian, A. (2009). Nutritional and antinutritional attributes of some under–utilized tubers. *Trop. Subtrop. Agroecosystems, 10*(2), 273-278.

Abdullahi Hassan, N., Karunakaran, R., Abdulmumin, S. (2019). A review on the pharmacological and traditional properties of *Mimosa pudica. Int. J. Pharm. Pharm. Sci., 11*(3), 12-16. [http://dx.doi.org/10.22159/ijpps.2019v11i3.30452]

Arjun, R.K., Asis, S., Jaleel, V.A., Sreena, R., Karthikeyan, S., Gothandam, K.M. (2012). Morphological, phytochemical, and anti-bacterial properties of wild and indigenous plant (*Amorphophallus commutatus*). *Acad. J., 7*(13), 744-748. [http://dx.doi.org/10.5897/JMPR12.269]

Arora, S., Meena, S. (2016). Morphological screening of endangered medicinal plants of milkweed family from Thar desert, Rajasthan, India. *Biosci. Biotechnol. Res. Commun., 9*(3), 406-414. [http://dx.doi.org/10.21786/bbrc/9.3/10]

Arun, C.H., Kumar, R.S., Srinu, S., Babu, G.L., Kumar, G.R., Babu, J.A. (2011). Anti-inflammatory activity of aqueous extract of leaves of *Solena amplexicaulis. Int. J. Res. Pharm. Biomed. Sci., 2*(4), 1617-1619.

Amirghofran, Z., Azadbakht, M., Karimi, M.H. (2000). Evaluation of the immunomodulatory effects of five herbal plants. *J. Ethnopharmacol., 72*(1-2), 167-172. [http://dx.doi.org/10.1016/S0378-8741(00)00234-8] [PMID: 10967468]

Arunachalam, K.D., Velmurugan, P., Raja, R.B. (2010). Anti-inflammatory and cytotoxic effects of extract from *Plumbago zeylanica. Afr. J. Microbiol. Res., 4*(12), 1239-1245.

Asha, V.V., Pushpangadan, P. (1999). Antipyretic activity of *Cardiospermum halicacabum. Indian J. Exp. Biol., 37*(4), 411-414. [PMID: 10641181]

Ashutosh, M., Kumar, A.A., Padhan, A.R. (2011). A Literature review on *Argyreia nervosa* (Burm. F.) Bojer. *Int. J. Res. Ayurveda Pharm., 2*(5), 1501-1504.

Ashwini, M., Lather, N., Bole, S., Vedamurthy, A.B., Balu, S. (2012). *In vitro* antioxidant and anti-inflammatory activity of *Coccinia grandis. Int. J. Pharm. Pharm. Sci., 4,* 239-242.

Asif, H.M., Rehman, S.U., Akram, M., Akhtar, N., Sultana, S., Rehman, J.U. (2014). Medicinal Properties of *Cucumis melo* Linn. *RADS J. Pharm. Allied Health Sci., 2*(1), 58-62.

Asiwe, J.A.N., Balane, A., Dacora, F.D. (2009). Evaluation of cowpea breeding lines for nitrogen fixation at ARC-Grain Crop Institute. Montana, USA: Potchefstroom. *South Africa,* 14-9.

Assad, R., Reshi, Z.A., Jan, S., Rashid, I. (2017). Biology of Amaranths. *Bot. Rev., 83*(4), 382-436. [http://dx.doi.org/10.1007/s12229-017-9194-1]

Asthana, J.G., Jain, S., Mishra, A., Vijaykant, M.S. (2001). Evaluation of antileprotic herbal drug combinations and their combination with Dapsone. *IDrugs, 38,* 82-86.

Atal, C.K., Kapur, B.M. (1982). Cultivation and Utilization of Medicinal Plants. Jamu-Tawi: Regional Research Laboratory. *CSIR, 548.*

Atkins, S. (2004). Verbenaceae.*Flowering Plants· Dicotyledons: Lamiales (except Acanthaceae including Avicenniaceae).* (pp. 449-468). Berlin, Heidelberg: Springer Berlin Heidelberg.
[http://dx.doi.org/10.1007/978-3-642-18617-2_25]

Attar, U.A., Ghane, S.G. (2017). Phytochemicals, antioxidant activity and phenolic profiling of *Diplocyclos palmatus* (L.) C. *Int. J. Pharm. Pharm. Sci., 9*(4), 101-106. b
[http://dx.doi.org/10.22159/ijpps.2017v9i4.16891]

Attar, U.A., Ghane, S.G. (2017). Proximate composition, antioxidant activities and phenolic composition of *Cucumis sativus* forma hardwickii (Royle) W. J. de Wilde & Duyfjes. *Int. J. Phytomed., 9*(1), 101-112. a
[http://dx.doi.org/10.5138/09750185.1922]

Austin, D.F. (1999). Caesar's weed (*Urena lobata*)—an invasive exotic or a Florida native. *Wildland Weeds, 3*(1), 13-16.

Aye, M.M., Aung, H.T., Sein, M.M., Armijos, C. (2019). A Review on the Phytochemistry, Medicinal Properties and Pharmacological Activities of 15 Selected Myanmar Medicinal Plants. *Molecules, 24*(2), 293.
[http://dx.doi.org/10.3390/molecules24020293] [PMID: 30650546]

Azad, A.K., Wan Azizi, W.S., Babar, Z.M., Labu, Z.K., Zabin, S. (2013). An overview on phytochemical, anti-inflammatory and anti-bacterial activity of *Basella alba* leaves extract. *J Sci Res., 14*, 650-655.
[http://dx.doi.org/10.5829/idosi.mejsr.2013.14.5.71225]

Acharya, R., Borkar, S.D., Naik, R., Shukla, V.J. (2015). Evaluation of phytochemical content, nutritional value and antioxidant activity of Phanji - *Rivea hypocrateriformis* (Desr.) Choisy leaf. *Ayu, 36*(3), 298-302.
[http://dx.doi.org/10.4103/0974-8520.182755] [PMID: 27313417]

Babu, A.V., Rao, R.S.C., Kumar, K.G., Babu, B.H., Satyanaray, P.V.V. (2009). Biological activity of *Merremia emarginata* crude extracts in different solvents. *Res. J. Med. Plant, 3*(4), 134-140.
[http://dx.doi.org/10.3923/rjmp.2009.134.140]

Babu, N.P., Pandikumar, P., Ignacimuthu, S. (2009). Anti-inflammatory activity of Albizia lebbeck Benth., an ethnomedicinal plant, in acute and chronic animal models of inflammation. *J. Ethnopharmacol., 125*(2), 356-360.
[http://dx.doi.org/10.1016/j.jep.2009.02.041] [PMID: 19643557]

Bose Mazumdar Ghosh, A., Banerjee, A., Chattopadhyay, S. (2022). An insight into the potent medicinal plant *Phyllanthus amarus* Schum. and Thonn. *Nucleus, 65*(3), 437-472.
[http://dx.doi.org/10.1007/s13237-022-00409-z] [PMID: 36407559]

Badwaik, H, Singh, MK, Thakur, D (2011). The Botany, Chemistry, Pharmacological and Therapeutic Application of *Oxalis corniculata* Linn. - A review. *Int J Phytomed, 3*, 1-8.

Bageel, A., Honda, M.D.H., Carrillo, J.T., Borthakur, D. (2020). Giant leucaena (*Leucaena leucocephala* subsp. glabrata): a versatile tree-legume for sustainable agroforestry. *Agrofor. Syst., 94*(1), 251-268.
[http://dx.doi.org/10.1007/s10457-019-00392-6]

Bagga, J. (2018). *Tamilnadia uliginosa* (Retz) Tirveng and Sastre;(Rubiaceae): a new distributional record with ethnomedicinal uses from Palamu division of Jharkhand. India. *Biosci. Disco., 9*(1), 97-99.

Bairaj, P., Nagarajan, S. (1982). Apigenin 7-O-glucuronide from the owers of *Asteracantha longifolia* Nees. *IDrugs, 19*, 150-152.

Bako, I.G., Mabrouk, M.A., Abubakar, A. (2009). Antioxidant effect of ethanolic seed extract of *Hibiscus sabdariffa* linn (Malvaceae) alleviate the toxicity induced by chronic administration of sodium nitrate on some haematological parameters in wistars rats. *Adv. J. Food Sci. Technol., 1*(1), 39-42.

Bal, T., Murthy, P.N., Sengupta, S. (2012). Isolation and analytical studies of mucilage obtained from the seeds of *Dillenia indica* (family Dilleniaceae) by use of various analytical techniques. *Asian J. Pharm. Clin. Res., 5*(3), 65-68.

Balan, A.P., Predeep, S.V. (2017). A taxonomic revision of the genus *Smithia* Ait. (Fabaceae) in South India. *Taiwania, 62*(2), 175.

Balkrishna, A, Anjali, S, Neelam, RD, Ashwini, S, Vedpriya, A (2023). Leea indica (Burm. f.) Merr.: A Systematized Acquaint. *World J Surg Surgical Res, 6*, 1468.

Bandyopadhyay, S., Mukherjee, S.K. (2009). Wild edible plants of Koch Bihar district. *Nat. Prod. Radiance, 8*(1), 64-72.

Banik, A., Nema, S., Shankar, D. (2014). Wild edible tuber and root plants available in bastarregion of Chhattisgarh. *Int J Crop Improv., 5*, 85-89.

Basu, B., Kirtikar, K. (1987). Indian Medicinal Plants *Dehradun: International Book Distributors, 2*(2).

Batisteli, A.F., Costa, R.O., Christianini, A.V. (2020). Seed abundance affects seed removal of an alien and a native tree in the Brazilian savanna: Implications for biotic resistance. *Austral Ecol., 45*(7), 1007-1015. [http://dx.doi.org/10.1111/aec.12922]

Behera, K.K., Bhuniya, B., Silva, J.A.T.D., Sahoo, S. (2010). Plantlets Regeneration of Potato Yam (*Dioscorea bulbifera* L.) Through *in vitro* Culture from Nodal Segments. *Int. J. Plant Biol., 4*(1), 37-41.

Behera, SK, Panda, A, Behera, SK, Misra, MK (2006). Medicinal Plants Used by the Kandhas of Kandhamal District of Orissa. *Ind J Tradit Knowl, 5*, 519-528.

Bello, Z.A., Walker, S. (2017). Evaluating AquaCrop model for simulating production of amaranthus (*Amaranthus cruentus*) a leafy vegetable, under irrigation and rainfed conditions. *Agric. For. Meteorol., 247*, 300-310. [http://dx.doi.org/10.1016/j.agrformet.2017.08.003]

Benthall AP. (1933), Dehradun: Thacker Spink and Co Calcutta. The Trees of Calcutta and its Neighbourhood; p. 513.

Berardini, N., Fezer, R., Conrad, J., Beifuss, U., Carle, R., Schieber, A. (2005). Screening of mango (*Mangifera indica* L.) cultivars for their contents of flavonol O- and xanthone C-glycosides, anthocyanins, and pectin. *J. Agric. Food Chem., 53*(5), 1563-1570. [http://dx.doi.org/10.1021/jf0484069] [PMID: 15740041]

Berger, A., Gremaud, G., Baumgartner, M., Rein, D., Monnard, I., Kratky, E., Geiger, W., Burri, J., Dionisi, F., Allan, M., Lambelet, P. (2003). Cholesterol-lowering properties of amaranth grain and oil in hamsters. *Int. J. Vitam. Nutr. Res., 73*(1), 39-47. [http://dx.doi.org/10.1024/0300-9831.73.1.39] [PMID: 12690910]

Bhadoriya, S.S., Ganeshpurkar, A., Narwaria, J., Rai, G., Jain, A.P. (2011). *Tamarindus indica* : Extent of explored potential. *Pharmacogn. Rev., 5*(9), 73-81. [http://dx.doi.org/10.4103/0973-7847.79102] [PMID: 22096321]

Bhalerao, S.A., Sharma, A.S. (2014). Ethenomedicinal, phytochemical and pharmacological profile of *Ficus religiosa* Roxb. *Int. J. Curr. Microbiol. Appl. Sci., 3*(11), 528-538.

Bhandari, M.M. (1990). *Flora of the Indian Desert*. (Revised Edition.). Jodhpur: Scientific Publishers.

Bhaskar, V.H., Balakrishnan, N. (2009). Analgesic, anti-inflammatory and antipyretic activities of *Pergularia daemia* and *Carissa carandas. Daru, 17*, 168-174.

Bhat, A.S., Menon, M.M. (1971). *Sesbania grandiflora* (a potential pulpwood). *Indian For., 97*(3), 128-144.

Bhattacharjee SK. (2001). Carica papaya, In: Hand book of Medicinal Plants, 3rd Revised Edn, by Shashi Jain (Ed), Pointer publisher, Jaipur; pp.1-71.

Bhattacharya, E., Mandal Biswas, S. (2022). First report of the hyperaccumulating potential of cadmium and lead by *Cleome rutidosperma* DC. with a brief insight into the chemical vocabulary of its roots. *Front. Environ. Sci., 10*, 830087.
[http://dx.doi.org/10.3389/fenvs.2022.830087]

Bhattacharya, S.K., Bhattacharya, A., Kumar, A., Ghosal, S. (2000). Antioxidant activity of *Bacopa monniera* in rat frontal cortex, striatum and hippocampus. *Phytother. Res., 14*(3), 174-179.
[http://dx.doi.org/10.1002/(SICI)1099-1573(200005)14:3<174::AID-PTR624>3.0.CO;2-O]   [PMID: 10815010]

Bigoniya P. (2012). Euphorbia latex: A magic potion or poison. V.K. Gupta (Ed.), Traditional and folk herbal medicine: recent researches, Daya Publishing House, New Delhi; 1.

Biren, N.S., Nayak, B.S., Bhatt, S.P., Jalalpure, S.S., Seth, A.K. (2007). The anti-inflammatory activity of the leaves of *Colocasia esculenta.* SPJ-. *Saudi Pharm. J., 15*(3-4), 228-232.

Biswas, M., Ghosh, P., Biswas, S., Dutta, A., Chatterjee, S. (2020). Phytochemical analysis and determination of *In vitro* antioxidant and antimicrobial activity of *Phyllanthus amarus* leaves extracts. *Int J Bot Stud., 5*(2), 483-490.

Berihun, T., Molla, E. (2017). Study on the diversity and use of wild edible plants in Bullen District Northwest Ethiopia. *J. Bot. (Egypt), 2017*, 1-10.
[http://dx.doi.org/10.1155/2017/8383468]

Biswas, N.R., Gupta, S.K., Das, G.K., Kumar, N., Mongre, P.K., Haldar, D., Beri, S. (2001). Evaluation of Ophthacare ® eye drops—a herbal formulation in the management of various ophthalmic disorders. *Phytother. Res., 15*(7), 618-620.
[http://dx.doi.org/10.1002/ptr.896] [PMID: 11746845]

Bitzer, J., Giraldi, A., Pfaus, J. (2013). Sexual desire and hypoactive sexual desire disorder in women. Introduction and overview. Standard operating procedure (SOP Part 1). *J. Sex. Med., 10*(1), 36-49.
[http://dx.doi.org/10.1111/j.1743-6109.2012.02818.x] [PMID: 22974089]

Block, G. (1992). The data support a role for antioxidants in reducing cancer risk. *Nutr. Rev., 50*(7), 207-213.
[http://dx.doi.org/10.1111/j.1753-4887.1992.tb01329.x] [PMID: 1641203]

Bolade, M.K., Oluwalana, I.B., Ojo, O. (2009). Commercial practice of roselle (*Hibiscus sabdariffa* L.) beverage production: Optimization of hot water extraction and sweetness level. *World J. Agric. Sci., 5*(1), 126-131.

Bole, P.V., Pathak, J.M. (1988). Flora of Saurashtra, Botanical Survey of India. *Part, III*, 392.

Boonmars, T., Khunkitti, W., Sithithaworn, P., Fujimaki, Y. (2005). *In vitro* antiparasitic activity of extracts of *Cardiospermum halicacabum* against third-stage larvae of *Strongyloides stercoralis. Parasitol. Res.,*

97(5), 417-419.
[http://dx.doi.org/10.1007/s00436-005-1470-z] [PMID: 16151739]

Boonyaprapas N, Chokchaijareonporn O. (1996). Samoon Prai Maipeunban (Herbal plants, in Thai) Thailand: Faculty of Pharmacy, Mahidol University.

Bora, N.S., Kakoti, B.B., Gogoi, B., Goswami, A.K. (2014). Ethno-medicinal claims, phytochemistry and pharmacology of *Spondias pinnata*: A review. *Int. J. Pharm. Sci. Rev. Res., 5*(4), 1138-45.
[http://dx.doi.org/10.13040/IJPSR.0975-8232.5]

Boskabady, M.H., Alitaneh, S., Alavinezhad, A. (2014). *Carum copticum* L.: a herbal medicine with various pharmacological effects. *BioMed Res. Int., 2014*, 1-11.
[http://dx.doi.org/10.1155/2014/569087] [PMID: 25089273]

Boulos, L. (1983). *Medicinal plants of North Africa.*

Boulos, L. (1999). *Flora of Egypt.* (p. 1). Cairo, Egypt: Al Hadara Publishing.

Brandbyge, J. (1993). Polygonaceae.*Flowering Plants· Dicotyledons: Magnoliid, Hamamelid and Caryophyllid Families.* (pp. 531-544). Berlin, Heidelberg: Springer Berlin Heidelberg.
[http://dx.doi.org/10.1007/978-3-662-02899-5_63]

Breitbach, U.B., Niehues, M., Lopes, N.P., Faria, J.E.Q., Brandão, M.G.L. (2013). Amazonian Brazilian medicinal plants described by C.F.P. von Martius in the 19th century. *J. Ethnopharmacol., 147*(1), 180-189.
[http://dx.doi.org/10.1016/j.jep.2013.02.030] [PMID: 23500885]

Brinkhaus, B., Lindner, M., Schuppan, D., Hahn, E.G. (2000). Chemical, pharmacological and clinical profile of the East Asian medical plant Centella aslatica. *Phytomedicine, 7*(5), 427-448.
[http://dx.doi.org/10.1016/S0944-7113(00)80065-3] [PMID: 11081995]

Britannica, T. Editors of Encyclopaedia (2022). Burseraceae. Encyclopedia Britannica. Available from: https://www.britannica.com/plant/Burseraceae.

Broderick, D.H. (1990). THE BIOLOGY OF CANADIAN WEEDS.: 93. *Epilobium angustifolium* L. (Onagraceae). *Can. J. Plant Sci., 70*(1), 247-259.
[http://dx.doi.org/10.4141/cjps90-027]

Brown, A.C., Reitzenstein, J.E., Liu, J., Jadus, M.R. (2005). The anti-cancer effects of poi(*Colocasia esculenta*) on colonic adenocarcinoma cells *In vitro*. *Phytother. Res., 19*(9), 767-771.
[http://dx.doi.org/10.1002/ptr.1712] [PMID: 16220568]

Brown AC, Valiere A. (2004). The medicinal uses of poi. Nutrition in clinical care: an official publication of Tufts University. 7(2): 69.

Brücher, H. (1989). *Useful plants of neotropical origin and their wild relatives..* Berlin, Germany: Springer Verlag.
[http://dx.doi.org/10.1007/978-3-642-73313-0]

Bruneton, J. (1999). Carica papaya.*Pharmacognosy, Phytochemistry of Medicinal plants.* (2nd ed., pp. 221-223). France: Technique & Documentation.

Bruneton, J. (1995). *Pharmacognosy, phytochemistry, medicinal plants..* Paris: Lavoisier Publishers, Translation of Pharmacognosie.

Buchwald, W., Kozłowski, J., Szczyglewska, D., Fory-cka, A. (2006). Biology of germination of medicinal plant seeds. Part XXII: Seeds of *Chamaenerion angustifolium* (L.) Scop. from Oenotheraceae family. *Herba Pol., 52*, 16-21.

Buerki, S., Callmander, M.W., Acevedo-Rodriguez, P., Lowry, P.P., II, Munzinger, J., Bailey, P., Maurin, O., Brewer, G.E., Epitawalage, N., Baker, W.J., Forest, F. (2021). An updated infra-familial classification of Sapindaceae based on targeted enrichment data. *Am. J. Bot., 108*(7), 1234-1251.
[http://dx.doi.org/10.1002/ajb2.1693] [PMID: 34219219]

Bulbul, I., Nahar, L., Haque, M. (2011). Antibacterial, cytotoxic and antioxidant activity of chloroform, n-hexane and ethyl acetate extract of plant Coccinia cordifolia. *Agric. Biol. J. N. Am., 2*(4), 713-719.
[http://dx.doi.org/10.5251/abjna.2011.2.4.713.719]

Buragohain, J. (2008). Folk medicinal plants used in gynecological disorders in Tinsukia district, Assam, India. *Fitoterapia, 79*(5), 388-392.
[http://dx.doi.org/10.1016/j.fitote.2008.03.004] [PMID: 18505703]

Burkill, H.M. (2004). The useful plants of West Tropical Africa. *Families S-Z, Addenda, Royal Botanic Gardens, Kew, Richmond, United Kingdom, 5*(2), 686.

Burkill, I.H. (1966). *A dictionary of the economic products of the Malay Peninsula.* (2nd ed.). London, England: Crown Agents.

Bylka, W., Kowalewski, Z. (1997). Flawonoidy w. *Chenopodium album* L. i *Chenopodium opulifolium* L. (Chenopodiaceae). *Herba Pol., 3*(43), 208-213.

Calalb, T., Fursenco, C., Chisnicean, L., Jelezneac, G., Balmuş, Z. (2021). Morphological and Anatomical Profile of (L.) Species Grown in the Republic of Moldova. *Acta Biologica Marisiensis, 5*(2), 1-8.
[http://dx.doi.org/10.2478/abmj-2022-0006]

Calixto, J.B., Santos, A.R.S., Filho, V.C., Yunes, R.A. (1998). A review of the plants of the genusPhyllanthus: Their chemistry, pharmacology, and therapeutic potential. *Med. Res. Rev., 18*(4), 225-258.
[http://dx.doi.org/10.1002/(SICI)1098-1128(199807)18:4<225::AID-MED2>3.0.CO;2-X] [PMID: 9664291]

Cartaxo, S.L., de Almeida Souza, M.M., de Albuquerque, U.P. (2010). Medicinal plants with bioprospecting potential used in semi-arid northeastern Brazil. *J. Ethnopharmacol., 131*(2), 326-342.
[http://dx.doi.org/10.1016/j.jep.2010.07.003] [PMID: 20621178]

Carvalho, F.A., Renner, S.S. (2013). The phylogeny of the Caricaceae.*Genetics and genomics of papaya.* (pp. 81-92). New York, NY: Springer New York.

Cesur, S., Demiröz, A.P. (2013). Antibiotics and the mechanisms of resistance to antibiotics. *Medical Journal of Islamic World Academy of Sciences, 21*(4), 138-142.
[http://dx.doi.org/10.12816/0002645]

Chae, S., Kang, K.A., Kim, J.S., Hyun, J.W., Kang, S.S. (2006). Trichotomoside: a new antioxidative phenylpropanoid glycoside from *Clerodendron trichotomum. Chem. Biodivers., 3*(1), 41-48.
[http://dx.doi.org/10.1002/cbdv.200690005] [PMID: 17193214]

Chaithra, D., Yasodamma, N., Alekhya, C. (2013). Phytochemical screening of *C. neilgherrensis* Wt. An endemic medicinal plant from Seshachalam Hills (A.P.) India. *Int. J. Pharma Bio Sci., 4*(2), 409-412.

Chakrabarty, T., Sarker, U., Hasan, M., Rahman, M.M. (2018). Variability in mineral compositions, yield and yield contributing traits of stem amaranth (*Amaranthus lividus*). *Genetika, 50*(3), 995-1010.
[http://dx.doi.org/10.2298/GENSR1803995C]

Chalil, S.S., Shinde, R.D. (2021). Wild edible leaves with medicinal uses found in Palghar district. *J. Glob Biosci., 10*(12), 9149-9181.

Chan K, Islam MW, Kamil MA, *et al.* (2000). The analgesic and anti-inflammatory effects of *Portulaca oleracea* L. subsp. sativa (Haw.) Celak. *J Ethnopharmacology.* 73(3): 445-51.
[http://dx.doi.org/10.1016/s0378-8741(00)00318-4]

Chandak, R.R., Dighe, N.S. (2019). A Review on Phytochemical & Pharmacological Profile of *Pergularia Daemia* linn. *J. Drug Deliv. Ther., 9*(4-s), 809-814.
[http://dx.doi.org/10.22270/jddt.v9i4-s.3426]

Chandrashekar, A., Adake, R., Rao, P.S., Santanusaha, S. (2013). *Wrightia Tinctoria*: An Overview. *J. Drug Deliv. Ther., 3*(2), 444.
[http://dx.doi.org/10.22270/jddt.v3i2.444]

Chase, M.W., Christenhusz, M.J., Fay, M.F. (2016). An update of the Angiosperm Phylogeny Group classification for the orders and families of flowering plants: APG IV. *Bot. J. Linn. Soc., 181*(1), 1-20.
[http://dx.doi.org/10.1111/boj.12385]

Chatterjee, A., Pakrashi, S.C. (2006). *The Treatise on Indian Medicinal Plants.* (pp. 65-66). New Delhi: National Institute of Science Commission and Information Resources.

Chatterjee, D. (1948). A review of Bignoniaceae of India and Burma. *Bulletin of the Botanical Society of Bengal, 2*, 62-79.

Chaudhuri, R.K., Ratan, K. (2004). Standardised extract of *Phyllanthus emblica*: A skin lightener with anti-aging benefits. *Proceedings PCIA Conference,* Guangzhou, China9-11.

Chauhan, N.S., Dixit, V.K. (2010). Effects of *Bryonia laciniosa* seeds on sexual behaviour of male rats. *Int. J. Impot. Res., 22*(3), 190-195.
[http://dx.doi.org/10.1038/ijir.2009.62] [PMID: 20032987]

Chavan, S., Dias, R., Magdum, C. (2021). *Garuga pinnata* attenuates oxidative stress and liver damage in chemically induced hepatotoxicity in rats. *Egypt. J. Basic Appl. Sci., 8*(1), 235-251.
[http://dx.doi.org/10.1080/2314808X.2021.1961207]

Chettleborough, J., Lumeta, J., Magesea, S. (2000). *Community use of nontimber forest products: A case study for the Kilombero valley. Integrated Environmental Programme. The Society for Environmental Exploration UK and the University of Da es Salaam.* Frontier Tazania.

Chhatre, S., Nesari, T., Kanchan, D., Somani, G., Sathaye, S. (2014). Phytopharmacological overview of *Tribulus terrestris. Pharmacogn. Rev., 8*(15), 45-51.
[http://dx.doi.org/10.4103/0973-7847.125530] [PMID: 24600195]

Chidambaram, K., Alqahtani, T., Alghazwani, Y., Aldahish, A., Annadurai, S., Venkatesan, K., Dhandapani, K., Thilagam, E., Venkatesan, K., Paulsamy, P., Vasudevan, R., Kandasamy, G. (2022). Medicinal Plants of *Solanum* Species: The Promising Sources of Phyto-Insecticidal Compounds. *J. Trop. Med., 2022*(1), 1-22.
[http://dx.doi.org/10.1155/2022/4952221] [PMID: 36187457]

Chinh, V.T., Quang, B.H., Anh, T.T. (2015). Morphological characteristics and key to genera of family Menispermaceae in Vietnam. *Proceedings of the 6th National Scientific Conference of Ecology and Biological Resources,* Hanoi, Vietnam 27-32.

Chiou, C.R., Wang, H.H., Chen, Y.J., Grant, W.E., Lu, M.L. (2013). Modeling potential range expansion of

the invasive shrub *Leucaena leucocephala* in the Hengchun peninsula, Taiwan. *Invasive Plant Sci. Manag.,* 6(4), 492-501.
[http://dx.doi.org/10.1614/IPSM-D-13-00010.1]

Chirumbolo, S. (2014). Dietary assumption of plant polyphenols and prevention of allergy. *Curr. Pharm. Des., 20*(6), 811-839.
[http://dx.doi.org/10.2174/13816128113199990042] [PMID: 23701556]

Chopra, R.N., Chopra, I.C., Varma, B.S. (1992). *Supplement to Glossary of Indian Medicinal Plants.* (reprinted edition., p. 29). New Delhi: CSIR.

Chopra, R.N., Nayer, S.L., Chopra, I.C. (1992). *Glossary of Indian medicinal plants.* (pp. 50-52). New Delhi: CSIR.

Chopra RN., Chopra IC., Handa KL., Kapur LD. (1958). Indigenous Drugs of India, U.N.Dhur and Sons Pvt. Ltd., Calcutta; 289, 665.

Chopra, S.R.R.N. (1984). *Poisonous Plants of India.* (Vol. 1, pp. 196-197). Jaipur.

Chopra, RN, Nayar, SL, and Chopra, IC, (1956). Glossary of Indian Medicinal plant, Council of scientific and industrial research; pp. 135-145.

Choudhury, S., Rahaman, C.H., Mandal, S., Ghosh, A. (2013). Folk-lore knowledge on medicinal usage of the tribal belts of Birbhum district, West Bengal, India. *Int. J. Bot. Res., 3*(2), 2277-4815.

Chowdhary, C.V., Meruva, A., K, N., Elumalai, R.K.A. (2013). A review on phytochemical and pharmacological profile of *Portulaca oleracea* Linn. (Purslane). *Int. J. Res. Ayurveda Pharm., 4*(1), 34-37.
[http://dx.doi.org/10.7897/2277-4343.04119]

Christenhusz, M.J.M., Byng, J.W. (2016). The number of known plants species in the world and its annual increase. *Phytotaxa, 261*(3), 201-217.
[http://dx.doi.org/10.11646/phytotaxa.261.3.1]

Chunekar, K.C., Pandey, G.S. (2006). *Guduchyadi Varga, Bhavprakash Nidhantu.* (p. 269). Varanasi: Chaukhambha Bharati Academy.

Conran, J.G. (1998). Smilacaceae.*Flowering Plants· Monocotyledons: Lilianae (except Orchidaceae).* (pp. 417-422). Berlin, Heidelberg: Springer Berlin Heidelberg.
[http://dx.doi.org/10.1007/978-3-662-03533-7_52]

Conti, M.V., Campanaro, A., Coccetti, P., De Giuseppe, R., Galimberti, A., Labra, M., Cena, H. (2019). Potential role of neglected and underutilized plant species in improving women's empowerment and nutrition in areas of sub-Saharan Africa. *Nutr. Rev., 77*(11), 817-828.
[http://dx.doi.org/10.1093/nutrit/nuz038] [PMID: 31313806]

Controller of Publications, Ministry of Health and Family Welfare, Department of Indian Systems of Medicine and Homoeopathy, Government of India, (2001).

Cook, C.D.K. (1996). *Aquatic and wetland plants of India..* New York: Oxford University press Inc.
[http://dx.doi.org/10.1093/oso/9780198548218.001.0001]

Cook. T, (1906). Flora of Presidency of Bombay, BSI, Calcutta; 2, p. 219.

Cooper, C.R., McLean, L., Walsh, M., Taylor, J., Hayasaka, S., Bhatia, J., Pienta, K.J. (2000). Preferential adhesion of prostate cancer cells to bone is mediated by binding to bone marrow endothelial cells as compared to extracellular matrix components in vitro. *Clin. Cancer Res., 6*(12), 4839-4847. [PMID: 11156243]

Council of Industrial and Scientific Research. The Wealth of India New Delhi: (2005); 3.

Craig, W.J. (1999). Health-promoting properties of common herbs. *Am. J. Clin. Nutr., 70*(3) (Suppl.), 491S-499S. [http://dx.doi.org/10.1093/ajcn/70.3.491s] [PMID: 10479221]

Czerwiński, J., Bartnikowska, E., Leontowicz, H., Lange, E., Leontowicz, M., Katrich, E., Trakhtenberg, S., Gorinstein, S. (2004). Oat (*Avena sativa* L.) and amaranth (*Amaranthus hypochondriacus*) meals positively affect plasma lipid profile in rats fed cholesterol-containing diets. *J. Nutr. Biochem., 15*(10), 622-629. [http://dx.doi.org/10.1016/j.jnutbio.2004.06.002] [PMID: 15542354]

Dagar, J.C., Singh, G., Singh, N.T. (1995). Evolution of crops in agroforestry with Teak (*Tectoma grandis*), Maharukh (*Ailanthus excelsa*) and Tamarind (*Tamarindus indica*) on reclaimed salt-affected soils. *J. Trop. For. Sci., 7*, 623-634.

Dajue, L., Mündel, H.H. (1996). *Safflower Carthamus tinctorius* L.. Italy: International Plant Genetic Resources Institute.

Dalziel, J.M. (1937). *The Useful Plants of West Tropical Africa.*. London: Crown Agents for the Colonies.

Dana E, Sanz-Elorza M, Sobrino E. Plant invaders in Spain (check-list). The unwanted citizens. Available from: http://www.med-alienplants.org/checklist. (2003).

Danish, M., Singh, P., Mishra, G., Srivastava, S., Jha, K.K., Khosa, R.L. (2011). *Cassia fistula* linn. (Amulthus)-an important medicinal plant: a review of its traditional uses, phytochemistry and pharmacological properties. *J Nat Prod Plant Resour., 1*, 101-118.

Darcansoy İşeri, Ö., Yurtcu, E., Sahin, F.I., Haberal, M. (2013). *Corchorus olitorius* (jute) extract induced cytotoxicity and genotoxicity on human multiple myeloma cells (ARH-77). *Pharm. Biol., 51*(6), 766-770. [http://dx.doi.org/10.3109/13880209.2013.765897] [PMID: 23577798]

Das, M, Kumar, A. (2013). Phyto-pharmacological review of *Portulaca quadrifida* Linn. *J. Appl. Pharm. Sci., 1*(1), 1-4.

Das PN, (2003). A Hand Book of Medicinal Plants, Jodpur: Agrobios.

Dashputre NL, Bandawane DD. (2021). Effect of *Abelmoschus ficulneus* (L.) Wight & Arn. on immunomodulation: *in vivo* experimental animal models. *Futur. J. Pharm. Sci.* 7: 1-1. [http://dx.doi.org/10.1186/s43094-021-00257-9]

Datar, M.N., Upadhye, A.S. (2016). *Forest foods of northern region of Western Ghats. Maharashtra Association for the Cultivation of Science (MACS).*. Pune: Agharkar Research Institute.

De Angelis, A., Gasco, L., Parisi, G., Danieli, P.P. (2021). A multipurpose leguminous plant for the Mediterranean countries: *Leucaena leucocephala* as an alternative protein source: a review. *Animals (Basel), 11*(8), 2230. [http://dx.doi.org/10.3390/ani11082230] [PMID: 34438688]

Debuigne, G., Couplan, F., Vignes, P., Vignes, D. (2009). *Petit Larousse des plantes médicinales.*. Paris: Larousse.

Deepak, Srivastav DS, Khare A. Progress in the Chemistry of Organic Natural Products. *Springerlink, 71*,

169-325.

Deepthy, M.J., Radhamany, M.M., Gayathridevi, V. (2016). Identification of phytochemical constituents from *Tamilnadia uliginosa* (RETZ.) Tirveng & Sastre (Rubiaceae) fruits using HPLC analysis. Definition of definitive cancer therapy by medical-dictionary. *World J. Pharm. Pharm. Sci., 5*, 826-830.

Demain, A.L., Fang, A. (2000). The natural functions of secondary metabolites. *Adv. Biochem. Eng. Biotechnol., 69*, 1-39.
[http://dx.doi.org/10.1007/3-540-44964-7_1] [PMID: 11036689]

Deokate, U.A., Khadabadi, S.S. (2011). Pharmacology and phytochemistry of *Coccinia indica. J. Pharmacogn. Phytother., 3*(11), 155-159.
[http://dx.doi.org/10.5897/JPP11.005]

Desh Bylka, S.V., Patil, M.J., Daswadhkar, S.C., Suralkar, U., Agarwal, A. (2011). A study on anti-inflammatory activity of the leaf and stem extracts of *Coccinia grandis* Voigt. *Int J Appl Bio Pharm., 2*, 247-250. b

Deshmukh, B.S., Waghmode, A. (2011). Role of wild edible fruits as a food resource: Traditional knowledge. *Int. J. Pharma Sci., 2*(7).

Deshmukh, S.A., Gaikwad, D.K. (2014). A review of the taxonomy, ethno botany, photochemistry and pharmacology of *Basella alba* (Basellaceae). *J. Appl. Pharm. Sci., 40*, 153-165.
[http://dx.doi.org/10.7324/JAPS.2014.40125]

Deshpande, S.V., Patil, M.J., Daswadhkar, S.C., Suralkar, U., Agarwal, A. (2011). A study on anti-inflammatory activity of the leaf and stem extracts of *Coccinia grandis* Voigt. *Int J Appl Bio Pharm., 2*, 247-250. b

Deshpande, S.M., Gosavi, K.V.C., Yadav, S.R. (2014). Karyomorphological Work in Two Endemic Species of Tricholepis (Asteraceae) in India. *Cytologia (Tokyo), 79*(4), 561-566.
[http://dx.doi.org/10.1508/cytologia.79.561]

Dev, S. (2006). *A Selection of Prime Ayurvedic Plant Drugs: Ancient-modern Concordance.* New Delhi: Anamaya.

Devi Prasad, A.G., Shyma, T.B. (2013). Medicinal plants used by the tribes of Vythiri taluk, Wayanad district (Kerala state) for the treatment of human and domestic animal ailments. *J. Med. Plants Res., 7*, 1439-1451.

Dey, Y., Srikanth, N., Wanjari, M., Ota, S., Jamal, M. (2012). A phytopharmacological review on an important medicinal plant - *Amorphophallus paeoniifolius. Ayu, 33*(1), 27-32.
[http://dx.doi.org/10.4103/0974-8520.100303] [PMID: 23049180]

Dhanalakshmi, R., Manavalan, R. (2014). *Bioactive compounds in leaves of Corchorus trilocularis L. by GC-MS analysis.* PharmTech.

Dhanasekaran, S., Muralidaran, P. (2010). CNS depressant and antiepileptic activities of the meth-anol extract of the leaves of *Ipomoea aquatica* Forsk. *E-J. Chem., 7*, 15-61.
[http://dx.doi.org/10.1155/2010/503923]

Dhanraj, M., Kadam, M., Patil, K., Mane, V. (2013). Phytochemical screening and antibacterial activity of western region wild leaf *Colocasia esculenta. Int. J. Biol. Sci., 2*(10), 18-23.

Dhawan, B.N., Dubey, M.P., Mehrotra, B.N., Rastogi, R.P., Tandon, J.S. (1980). Screening of Indian plants for biological activity: Part IX. *Indian J. Exp. Biol., 18*(6), 594-606.

[PMID: 7439945]

Dhodade, P.N., Dhaygude, Y.P., Tiwari, A.K., Birwatkar, V.R. (2019). Development of Jam from Under Exploited Fruit Aliv (*Meyna laxiflora* Robyns). *Int. J. Curr. Microbiol. Appl. Sci., 8*(3), 1143-1152. [http://dx.doi.org/10.20546/ijcmas.2019.803.136]

Dilip, K.J., Saroja, K., Murthy, A.R. (2012). Pharmacognostic study of Kakamachi (*Solanum nigrum* Linn). *J. Pharm. Innov., 1*(4), 42-48.

Diaga, D. (2011). Recent advances in cowpea [*Vigna unguiculata* (L.) Walp.] omics research for genetic improvement. *Afr. J. Biotechnol., 10*(15), 2803-2819. [http://dx.doi.org/10.5897/AJBx10.015]

Dkhil, M.A., Moniem, A.E.A., Al-Quraishy, S., Saleh, R.A. (2011). Antioxidant effect of purslane (*Portulaca oleracea*) and its mechanism of action. *J. Med. Plants Res., 5*(9), 1589-1593.

Doblado, R., Zielinski, H., Piskula, M., Kozlowska, H., Muñoz, R., Frías, J., Vidal-Valverde, C. (2005). Effect of processing on the antioxidant vitamins and antioxidant capacity of *Vigna sinensis* Var. Carilla. *J. Agric. Food Chem., 53*(4), 1215-1222. [http://dx.doi.org/10.1021/jf0492971] [PMID: 15713044]

Dohnal, B. (2015). Investigations on some metabolites of *Tecoma stans* Juss. callus tissue. Part III. Chromatographical search for iridoids, phenolic acids, terpenoids and sugars. *Acta Soc. Bot. Pol., 46*(2), 187-199. [http://dx.doi.org/10.5586/asbp.1977.015]

dos Santos Magalhães, C., dos Santos Melo, D.F., da Silva, H.C.C., de Carvalho, R.R., da Silva, R.V.L., de Caldas Brandão Filho, J.O., da Silva, F.C.L., Randau, K.P. (2023). A Review of the Ethnobotany, Phytochemistry, and Pharmacology of the Family Cleomaceae of Brazilian Origin. *J. Herb. Med., 42*, 100814. [http://dx.doi.org/10.1016/j.hermed.2023.100814]

Dua, V.K., Verma, G., Singh, B., Rajan, A., Bagai, U., Agarwal, D.D., Gupta, N.C., Kumar, S., Rastogi, A. (2013). Anti-malarial property of steroidal alkaloid conessine isolated from the bark of *Holarrhena antidysenterica*. *Malar. J., 12*(1), 194. [http://dx.doi.org/10.1186/1475-2875-12-194] [PMID: 23758861]

Duan, W., Yu, Y., Zhang, L. (2005). Antiatherogenic effects of *Phyllanthus emblica* associated with corilagin and its analogue. *Yakugaku Zasshi, 125*(7), 587-591. [http://dx.doi.org/10.1248/yakushi.125.587] [PMID: 15997216]

Dugje, I.Y., Ekeleme, F., Kamara, A.Y. (2008). Guide to safe and effective use of pesticides for crop production. *International Institute of Tropical Agriculture, Ibadan.* [http://dx.doi.org/10.13140/2.1.2721.8566]

Dugje, I.Y., Omoigui, L.O., Ekeleme, F., Kamara, A.Y., Ajeigbe, H. (2009). *Production du niébé en Afrique de l'Ouest: Guide du paysan.*. Ibadan, Nigéria: IITA. Internet

Duke, J.A., Ayensu, E.S. (1985). Medicinal Plants of China, 1300 Strichzeichnungen. Reference Publ. *Inc. Algonac. Michigan. USA, 2*(705), 17-25.

Etyemez Büyükdeveci, M., Balcázar, J.L., Demirkale, İ., Dikel, S. (2018). Effects of garlic-supplemented diet on growth performance and intestinal microbiota of rainbow trout (Oncorhynchus mykiss). *Aquaculture, 486*, 170-174.

[http://dx.doi.org/10.1016/j.aquaculture.2017.12.022]

Edeoga, H.O., Okwu, D.E., Mbaebie, B.O. (2005). Phytochemical constituents of some Nigerian medicinal plants. *Afr. J. Biotechnol., 4*(7), 685-688.
[http://dx.doi.org/10.5897/AJB2005.000-3127]

Editorial Board of Chinese Materia Medica Huangyaozi. (1999) Zhong Hua Ben Cao (Chinese Materia Medica) Shanghai Science and Technology Press, Shanghai; 8 pp. 7278-7280.

Eggeling, W.J., Dale, I.R. (1951). *The Indigenous Trees of the Uganda Protectorate Entebbe.* (p. 491). Uganda: The Government Printer.

Ekanayake, S., Jansz, E.R., Abeysekera, A.M., Nair, B.M. (2001). Some anti-nutritional factors of mature sword beans (*Canavalia gladiata*). *Vidyodaya J. Sci., 10*, 81-90.

El Gendy, A.N.G., Tavarini, S., Conte, G., Pistelli, L., Hendawy, S.F., Omer, E.A., Angelini, L.G. (2017). Yield and qualitative characterisation of seeds of *Amaranthus hypochondriacus* L. and *Amaranthus cruentus* L. grown in central Italy. *Ital. J. Agron., 13*(1), 63-73.
[http://dx.doi.org/10.4081/ija.2017.993]

El-Desouky, S.K., Ryu, S.Y., Kim, Y.K. (2008). A new cytotoxic acylated apigenin glucoside from *Phyllanthus emblica* L. *Nat. Prod. Res., 22*(1), 91-95.
[http://dx.doi.org/10.1080/14786410701590236] [PMID: 17999342]

Erarslan, Z.B., Koçyiğit, M. (2019). The important taxonomic characteristics of the family Malvaceae and the herbarium specimens in ISTE. *Turkish Journal of Bioscience and Collections, 3*(1), 1-7.
[http://dx.doi.org/10.26650/tjbc.20190001]

Evans SR, Frust PT. (1990). An Overview of Hallucinogens: The Flash of God, Waveland Press, Long Grove, ILUSA; Vol. 3–54.

Faden, R.B. (1998). Commelinaceae.*Flowering Plants· Monocotyledons: Alismatanae and Commelinanae (except Gramineae).* (pp. 109-128). Berlin, Heidelberg: Springer Berlin Heidelberg.
[http://dx.doi.org/10.1007/978-3-662-03531-3_12]

Fatima, M., Ahmed, S., Siddiqui, M.U. ul Hasan MM (2021). Medicinal uses, phytochemistry and pharmacology of *Bauhinia racemosa* lam. *J. Phcog. Phytochem., 10*(2), 121-124.
[http://dx.doi.org/10.22271/phyto.2021.v10.i2b.13972]

Fauré, M., Lissi, E., Torres, R., Videla, L.A. (1990). Antioxidant activities of lignans and flavonoids. *Phytochemistry, 29*(12), 3773-3775.
[http://dx.doi.org/10.1016/0031-9422(90)85329-E]

Feldman, K.S. (2005). Recent progress in ellagitannin chemistry. *Phytochemistry, 66*(17), 1984-2000.
[http://dx.doi.org/10.1016/j.phytochem.2004.11.015] [PMID: 16153404]

Fergany, M., Kaur, B., Monforte, A.J., Pitrat, M., Rys, C., Lecoq, H., Dhillon, N.P.S., Dhaliwal, S.S. (2011). Variation in melon (*Cucumis melo*) landraces adapted to the humid tropics of southern India. *Genet. Resour. Crop Evol., 58*(2), 225-243.
[http://dx.doi.org/10.1007/s10722-010-9564-6]

Ferreira Ozela, E., Stringheta, P.C., Cano Chauca, M. (2007). Stability of anthocyanin in spinach vine (*Basella rubra*) fruits. *Cienc. Investig. Agrar., 34*(2), 115-120.
[http://dx.doi.org/10.4067/S0718-16202007000200004]

Fischer, E., Theisen, I., Lohmann, L.G. (2004). Bignoniaceae.*Flowering Plants· Dicotyledons: Lamiales (except Acanthaceae including Avicenniaceae).* (pp. 9-38). Berlin, Heidelberg: Springer Berlin Heidelberg. [http://dx.doi.org/10.1007/978-3-642-18617-2_2]

Flora of China Editorial Committee. (2015). *Flora of China..* St. Louis, Missouri and Cambridge, Massachusetts, USA: Missouri Botanical Garden and Harvard University Herbaria.

Flora of North America Editorial Committee. (2015). *Flora of North America North of Mexico..* St. Louis, Missouri and Cambridge, Massachusetts, USA: Missouri Botanical Garden and Harvard University Herbaria.

Florence, PS (2019). *Use of Factory Statistics in the Investigation of Industrial Fatigue.,* 56-60.

Foster, S., Duke, J.A. (2000). *A field guide to medicinal plants and herbs of eastern and central North America..* Houghton Mifflin Harcourt.

Francis JK, (2004). Tropical ecosystems, *Ficus* spp. (and other important Moraceae), Encyclopedia of Forest Sciences; 1699-1704.

Fujii, T., Wakaizumi, M., Ikami, T., Saito, M. (2008). Amla (*Emblica officinalis* Gaertn.) extract promotes procollagen production and inhibits matrix metalloproteinase-1 in human skin fibroblasts. *J. Ethnopharmacol., 119*(1), 53-57. [http://dx.doi.org/10.1016/j.jep.2008.05.039] [PMID: 18588964]

Gabel, N.H., Wise, R.R., Rogers, G.K. (2020). Distribution of cystoliths in the leaves of Acanthaceae and its effect on leaf surface anatomy. *Blumea, 65*(3), 224-232.

Gai Guo, L.Y., Fan, S., Jing, S., Yan, L.J. (2016). Antioxidant and Antiproliferative Activities of Purslane Seed Oil. *J. Hypertens., 5*(2) [http://dx.doi.org/10.4172/2167-1095.1000218]

Gupta, H., Singhal, A.K., Bhati, V.S. (2011). Wound healing activity of *Argyreia nervosa* leaves extract. *Int. J. Appl. Basic Med. Res., 1*(1), 36-39. [http://dx.doi.org/10.4103/2229-516X.81978] [PMID: 23776770]

Gauthaman, K., Ganesan, A.P. (2008). The hormonal effects of *Tribulus terrestris* and its role in the management of male erectile dysfunction – an evaluation using primates, rabbit and rat. *Phytomedicine, 15*(1-2), 44-54. [http://dx.doi.org/10.1016/j.phymed.2007.11.011] [PMID: 18068966]

Galani, V. (2019). Musa paradisiaca. Linn. A Comprehensive Review. *Sch Int J Tradit Complement Med.,* 45-56.

Ganapathy, P.S., Ramachandra, Y.L., Rai, S.P. (2011). *In vitro* antioxidant activity of *Holarrhena antidysenterica* Wall. methanolic leaf extract. *J. Basic Clin. Pharm., 2*(4), 175-178. [PMID: 24826020]

Ganbold, M., Shimamoto, Y., Ferdousi, F., Tominaga, K., Isoda, H. (2019). Antifibrotic effect of methylated quercetin derivatives on TGFβ-induced hepatic stellate cells. *Biochem. Biophys. Rep., 20*, 100678. [http://dx.doi.org/10.1016/j.bbrep.2019.100678] [PMID: 31467991]

Gang, W., Binbin, L., Jinsong, L., Guokai, W., Fei, W., Jikai, L. (2009). Chemical constituents from tubers of *Dioscorea bulbifera*. China J. *Chinese Materia Medica, 34*(13), 1679-1682. [PMID: 19873780]

Gangopadhyay, M., Das, A.K., Bandyopadhyay, S., Das, S. (2021). Water Spinach (*Ipomoea aquatica* Forsk.) Breeding. Advances in Plant Breeding Strategies: Vegetable Crops: Leaves, Flowerheads, Green Pods. *Mushrooms and Truffles, 10*, 183-215.

Gantait, A., Barman, T., Mukherjee, P.K. (2011). Validated method for estimation of curcumin in turmeric powder. *Indian J. Tradit. Knowl., 10*(2), 247-250.

Gantait, S., Das, A., Mitra, M., Chen, J.T. (2021). Secondary metabolites in orchids: Biosynthesis, medicinal uses, and biotechnology. *S. Afr. J. Bot., 139*, 338-351.
[http://dx.doi.org/10.1016/j.sajb.2021.03.015]

Garodia, P., Ichikawa, H., Malani, N., Sethi, G., Aggarwal, B.B. (2007). From ancient medicine to modern medicine: ayurvedic concepts of health and their role in inflammation and cancer. *J. Soc. Integr. Oncol., 5*(1), 25-37.
[http://dx.doi.org/10.2310/7200.2006.029] [PMID: 17309811]

Gautam, S., Bhagyawant, S.S., Srivastava, N. (2014). Detailed study on therapeutic properties, uses and pharmacological applications of safflower (*Carthamus tinctorius* L.). *Int. J. Ayurveda Pharma Res., 2*(3), 1-2.

Gautam, V.P., Aslam, P.R., Bharti, K.U., Singhai, A.K. (2013). *Diplocyclos palmatus*: a phytopharmacological review. *Int. J. Pharm. Res., 3*(1), 157-159.

Gawade, B, Kode, N (2020). *Sidhudurgatil ranbhajya, published by Pandit Publication., 23*.

George CK, Rao YS. (1997). Tirupathi. India (A.P.) Export of Tamarind from India, Proceedings of National Symposium on *Tamarindus indica* L. organized by Forest Dept. of A.P., India; 27-28.

Ghaffar, A., Tung, B.T., Rahman, R., Nadeem, F., Idrees, M. (2019). Botanical Specifications, Chemical Composition and Pharmacological Applications of Tartara (*Digera muricata* L.)–A Review. *Int. J. Chem. Biochem. Sci., 16*, 17-22.

Ghani, A. (2003). Medicinal plants of Bangladesh: chemical constituents and uses. *Asiatic society of Bangladesh.*

Ghazanfar SA. (1994). Handbook of Arabian medicinal plants. CRC press; 24.
[http://dx.doi.org/10.1201/b14834]

Ghosh, P., Chatterjee, S., Das, P., Karmakar, S., Mahapatra, S. (2019). Natural habitat, phytochemistry and pharmacological properties of a medicinal weed *Cleome Rutidosperma* DC. (Cleomaceae): A comprehensive review. *Int. J. Pharma Sci., 10*(4), 1605-1612.
[http://dx.doi.org/10.13040/IJPSR.0975-8232.10(4).1605-12]

Gollucke, A.P.B., Aguiar, O., Jr, Barbisan, L.F., Ribeiro, D.A. (2013). Use of grape polyphenols against carcinogenesis: putative molecular mechanisms of action using *in vitro* and *in vivo* test systems. *J. Med. Food, 16*(3), 199-205.
[http://dx.doi.org/10.1089/jmf.2012.0170] [PMID: 23477622]

Gomase, P., Gomase, P., Anjum, S., Shakil, S., Shahnavaj, K.M. (2012). *Sesbania sesban* Linn: a review on its ethnobotany, phytochemical and pharmacological profile. *Asian J Biomed Pharm Sci., 2*(12), 11-14.

Ghosh, S., Saha, S. (2012). *Tinospora cordifolia*: One plant, many roles. *Anc. Sci. Life, 31*(4), 151-159.
[http://dx.doi.org/10.4103/0257-7941.107344] [PMID: 23661861]

Gonçalves, R.F., Silva, A.M.S., Silva, A.M., Valentão, P., Ferreres, F., Gil-Izquierdo, A., Silva, J.B., Santos, D., Andrade, P.B. (2013). Influence of taro (Colocasia esculenta L. Shott) growth conditions on the phenolic composition and biological properties. *Food Chem., 141*(4), 3480-3485.
[http://dx.doi.org/10.1016/j.foodchem.2013.06.009] [PMID: 23993510]

González-García, E., Cáceres, O., Archimède, H., Santana, H. (2009). Nutritive value of edible forage from two *Leucaena leucocephala* cultivars with different growth habit and morphology. *Agrofor. Syst., 77*(2), 131-141.
[http://dx.doi.org/10.1007/s10457-008-9188-4]

Govindachari, T.R., Nagarajan, K., Pai, B.R. (1957). Isolation of lupeol from the root of *Asteracantha longifolia* Nees. *J Indian Sci Res, 16B*, 72.

Gray, N.E., Alcazar Magana, A., Lak, P., Wright, K.M., Quinn, J., Stevens, J.F., Maier, C.S., Soumyanath, A. (2018). *Centella asiatica*: phytochemistry and mechanisms of neuroprotection and cognitive enhancement. *Phytochem. Rev., 17*(1), 161-194.
[http://dx.doi.org/10.1007/s11101-017-9528-y] [PMID: 31736679]

Green, P.S. (2004). Oleaceae.*Flowering Plants· Dicotyledons: Lamiales (except Acanthaceae including Avicenniaceae).* (pp. 296-306). Berlin, Heidelberg: Springer Berlin Heidelberg.
[http://dx.doi.org/10.1007/978-3-642-18617-2_16]

Grieve M. Purslane, (1995). Golden A modern Herbal homepage. botanical. Com.

Grover, J.K., Adiga, G., Vats, V., Rathi, S.S. (2001). Extracts of *Benincasa hispida* prevent development of experimental ulcers. *J. Ethnopharmacol., 78*(2-3), 159-164.
[http://dx.doi.org/10.1016/S0378-8741(01)00334-8] [PMID: 11694361]

Grubben GJ, Denton OA. (2004). Plant resources of tropical Africa 2. Vegetable. Wageningen, Leiden: PROTA Foundation, Backhuys, CTA; 103-111.

Grubben GJH. *Amaranthus cruentus* L. In: Grubben GJH, Denton OA, (2004) eds. PROTA (Plant Resources of Tropical Africa. Vegetables/Resources végétales de l'Afrique tropicale. Légumes. Wageningen, The Netherlands: Fondation PROTA, Backhuys Publishers; 2: 73–79.

Guan, X.R., Zhu, L., Xiao, Z.G., Zhang, Y.L., Chen, H.B., Yi, T. (2017). Bioactivity, toxicity and detoxification assessment of *Dioscorea bulbifera* L.: a comprehensive review. *Phytochem. Rev., 16*(3), 573-601.
[http://dx.doi.org/10.1007/s11101-017-9505-5]

Gucker C L. *Dioscorea* spp. (2009). In: Fire Effects Information System, U.S. Department of Agriculture, Forest Service, Rocky Mountain Research Station, Fire Sciences Laboratory, US.

Guhabakshi, D.N., Sensarma, P., Pal, D.C. (1999). *A lexicon of medicinal Plants in India.* (p. 1). New Delhi.

Gul, K., Singh, A.K., Jabeen, R. (2016). Nutraceuticals and functional foods: the foods for the future world. *Crit. Rev. Food Sci. Nutr., 56*(16), 2617-2627.
[http://dx.doi.org/10.1080/10408398.2014.903384] [PMID: 25629711]

Gupta, N., Sharma, D., Rani, R. (2022). *Bombax ceiba* LINN: A Critical review on phytochemistry, traditional uses, pharmacology, and toxicity from phytopharmaceutical perspective. *Int. J. Pharm. Pharm. Sci., 15*(1), 8-15.

Gupta, N., Shrivastava, N., Singh, P.K., Bhagyawant, S.S. (2016). Phytochemical evaluation of moth bean

(*Vigna aconitifolia* L.) seeds and their divergence. *Biochem. Res. Int., 2016*, 1-6. [http://dx.doi.org/10.1155/2016/3136043] [PMID: 27239343]

Gupta, R.C. (1997). Botanical identity of Jivanti the ayurevedic rejuvenant par excellence. *Appl. Bot. Abstr., 17*(1), 49-63.

Gupta RK. (2010). Medicinal and aromatic plants. CBS Publ Distrib 234:499 Return.

Gate, N. (2013). *Asparagus officinalis.*www.luontoportti.com/suomi/en/kukkakasvit /asparagus

Gupta, S., Kumar, J., Mishra, P., Singh, A.K., Maral, A. (2023). Morphological and pharmacological study of herbal medicine: *Corchorus olitorius* L. *J. Pharmacogn. Phytochem., 12*(2), 84-87.

Gupta, S., Jyothi Lakshmi, A., Manjunath, M.N., Prakash, J. (2005). Analysis of nutrient and antinutrient content of underutilized green leafy vegetables. *Lebensm. Wiss. Technol., 38*(4), 339-345. [http://dx.doi.org/10.1016/j.lwt.2004.06.012]

Gurrola-Díaz, C.M., García-López, P.M., Sánchez-Enríquez, S., Troyo-Sanromán, R., Andrade-González, I., Gómez-Leyva, J.F. (2010). Effects of *Hibiscus sabdariffa* extract powder and preventive treatment (diet) on the lipid profiles of patients with metabolic syndrome (MeSy). *Phytomedicine, 17*(7), 500-505. [http://dx.doi.org/10.1016/j.phymed.2009.10.014] [PMID: 19962289]

Habib-ur-Rehman, , Yasin, K.A., Choudhary, M.A., Khaliq, N., Atta-ur-Rahman, , Choudhary, M.I., Malik, S. (2007). Studies on the chemical constituents of *Phyllanthus emblica. Nat. Prod. Res., 21*(9), 775-781. [http://dx.doi.org/10.1080/14786410601124664] [PMID: 17763100]

Harborne, J.B. (1993). Major medicinal plants of India : edited by R. S. Thakur, H. S. Puri and A. Husain, Central Institute of Medicinal and Aromatic Plants, Lucknow, India, 1989. 585 pp. $150 (inc. postage). *Phytochemistry*, 33.

Hafaz, M.F., Soliman, H.M., Abbas, M.A., Gebreil, A.S., El-Amier, Y.A. (2022). Potential Assessment of Rumex spp. as a source of bioactive compounds and biological activity. *Biointerface Res. Appl. Chem., 12*, 1824-1834.

Hajlaoui, H., Arraouadi, S., Mighri, H., Ghannay, S., Aouadi, K., Adnan, M., Elasbali, A.M., Noumi, E., Snoussi, M., Kadri, A. (2022). HPLC-MS profiling, antioxidant, antimicrobial, antidiabetic, and cytotoxicity activities of *Arthrocnemum indicum* (Willd.) Moq. extracts. *Plants, 11*(2), 232. [http://dx.doi.org/10.3390/plants11020232] [PMID: 35050120]

Hall, A.E., Cisse, N., Thiaw, S., Elawad, H.O.A., Ehlers, J.D., Ismail, A.M., Fery, R.L., Roberts, P.A., Kitch, L.W., Murdock, L.L., Boukar, O., Phillips, R.D., McWatters, K.H. (2003). Development of cowpea cultivars and germplasm by the Bean/Cowpea CRSP. *Field Crops Res., 82*(2-3), 103-134. [http://dx.doi.org/10.1016/S0378-4290(03)00033-9]

Hall DW, Vernon VV, Brent A. Sellers. (1996). Creeping Wood Sorrel, *Oxalis corniculata* L. Southern Yellow Wood Sorrel, Oxalis florida Salisb. SP 37, Florida Cooperative Extension Service, Institute of Food and Agricultural Sciences, University of Florida, 01-02.

Hamed, K.A., Hassan, S.A., Mohamed, A., Hosney, N.K. (2014). Morphological and anatomical study on Plantaginaceae Juss. and some related taxa of Scrophulariaceae Juss. *Egypt. J. Exp. Biol. (Bot.), 10*(2), 135-146.

Hammer K. Aizoaceae. In: Hanelt P, (1986). Institute of Plant Genetics and Crop Plant Research (eds) Mansfeld'sencyclopedia on agricultural and horticultural crops. Springer Verlag, Berlin, Heidelberg, New York 2001; 1, pp 223–227.

Haque, M.A. (1996). *Agroforestry in Bangladesh. Village and Farm Forestry Project, SDC. Dhaka and BAU.* Mymensingh.

Harborne, J.B. (1973). *Phytochemical methods..* London: Chapman and Hall.

Harrison D, Harrison E, (2003) inventors. Natural therapeutic composition for the treatment of wounds and sores. United States patent application US 10/188,828.

Hasan, S.M.R., Hossain, M.M., Akter, R., Jamila, M., Mazumder, M.E.H., Alam, M.A., Faruque, A., Rana, S., Rahman, S. (2009). Analgesic activity of the different fractions of the aerial parts of *Commelina benghalensis. Int. J. Pharmacol., 6*(1), 63-67.
[http://dx.doi.org/10.3923/ijp.2010.63.67]

Hassan, S.T.S., Berchová, K., Šudomová, M. (2016). Antimicrobial, antiparasitic and anticancer properties of *Hibiscus sabdariffa* (L.) and its phytochemicals: *in vitro* and *in vivo* studies. *Ceska Slov. Farm., 65*(1), 10-14. [PMID: 27118499]

Heyde, H. (1990). *Medicijn planten in Suriname (Den dresi wiwiri foe Sranan). Medicinal Plants in Suriname. Uitg. Stichting Gezondheidsplanten Informaite.* (p. 157). Paramaribo: SGI.

Hishika, R., Shastry, S., Shinde, S., Guptal, S.S. (1981). Preliminary phytochemical and anti-inflammatory activity of seeds of *Mucuna pruriens. Indian J. Pharmacol., 13*(1), 97-98.

Hoang, L.S.H., Phan, T.H.Y. (2014). Preliminary Phytochemical Screening, Acute Oral Toxicity and Anti-convulsant Activity of the Barries of *Solanum nigrum* Linn. *Trop. J. Pharm. Res., 13*(6), 4.
[http://dx.doi.org/10.4314/tjpr.v13i6.12]

Holderness, J., Hedges, J.F., Daughenbaugh, K., Kimmel, E., Graff, J., Freedman, B., Jutila, M.A. (2008). Response of gammadelta T Cells to plant-derived tannins. *Crit. Rev. Immunol., 28*(5), 377-402.
[http://dx.doi.org/10.1615/CritRevImmunol.v28.i5.20] [PMID: 19166386]

Holm, L.G., Plucknett, D.L., Pancho, J.V., Herberger, J.P. (1991). *The world's worst weeds..* Malabar, FL: Kriegar Publishing Company.

Hong, D., DeFillipps, R.A. (2000). Commelina diffusa.*Flora of China, Beijing Science Press.* (p. 86). St Louis: Missouri Botanical Garden Press.

Hopkins, C.Y., Ewing, D.F., Chisholm, M.J. (1968). A short-chain ester from the seed oil of *Cardiospermum halicacabum* L. *Phytochemistry, 7*(4), 619-624.
[http://dx.doi.org/10.1016/S0031-9422(00)88237-0]

Horvath F, (1992) inventor; Unipharma Co Ltd, assignee. Therapeutical compositions against psoriasis. United States patent US 5,165,932.

Hossain, F., Saha, S., Islan, M.M., Nasrin, S., Adhikari, S. (2014). Analgesic and anti-inflammatory activity of *Commelina benghalensis* Linn. *Turk J Pharm Set, 11*(1), 25-32.

Hossain, M.S., Jahan, I., Islam, M., Nayeem, J., Anzum, T.S., Afrin, N.A., Mim, F.K., Hasan, M.K. (2024). *Coccinia grandis*: Phytochemistry, pharmacology and health benefits. *Clinical Traditional Medicine and Pharmacology, 5*(2), 200150.
[http://dx.doi.org/10.1016/j.ctmp.2024.200150]

Huang, C.C., Lin, K.J., Cheng, Y.W., Hsu, C.A., Yang, S.S., Shyur, L.F. (2013). Hepatoprotective effect and mechanistic insights of deoxyelephantopin, a phyto-sesquiterpene lactone, against fulminant hepatitis. *J.*

Nutr. Biochem., 24(3), 516-530.
[http://dx.doi.org/10.1016/j.jnutbio.2012.01.013] [PMID: 22748804]

Huang, X., Kong, L. (2006). Steroidal saponins from roots of *Asparagus officinalis. Steroids, 71*(2), 171-176.
[http://dx.doi.org/10.1016/j.steroids.2005.09.005] [PMID: 16280142]

Huang, Y.L., Oppong, M.B., Guo, Y., Wang, L.Z., Fang, S.M., Deng, Y.R., Gao, X.M. (2019). The Oleaceae family: A source of secoiridoids with multiple biological activities. *Fitoterapia, 136*, 104155.
[http://dx.doi.org/10.1016/j.fitote.2019.04.010] [PMID: 31028819]

Husain, A., Virmani, O.P., Popli, S.P. (1992). *Dictionary of Indian medicinal plants.* (p. 546). Lucknow, India: Central Institue of Medicinal and Aromatic Plants.

Hussain, A.Z., Kumaresan, S. (2014). GC-MS studies and phytochemical screening of *Sesbania grandiflora. J. Chem. Pharm. Res., 6*, 43-47.

Hussain, M.S., Fareed, S., Ali, M. (2010). *Hygrophila auriculata* (K. Schum) Heine: Ethnobotany, phytochemistry and pharmacology. *Asian J. Tradit. Med., 5*(4), 121-131.

Hussain, N., Kakoti, B.B. (2013). Review on ethnobotany and psychopharmacology of *Cordia dichotoma. J. Drug Deliv. Ther., 3*(1), 110-113.
[http://dx.doi.org/10.22270/jddt.v3i1.386]

Hussain, Z., Amresh, G., Singh, S., Rao, C.V. (2009). Antidiarrheal and antiulcer activity of *Amaranthus spinosus* in experimental animals. *Pharm. Biol., 47*(10), 932-939.
[http://dx.doi.org/10.1080/13880200902950769]

Hwang, C.Y., Hsu, L.M., Liou, Y.J., Wang, C.Y. (2010). Distribution, growth, and seed germination ability of lead tree (*Leucaena leucocephala*) plants in Penghu Islands, Taiwan. *Weed Technol., 24*(4), 574-582.
[http://dx.doi.org/10.1614/WT-D-10-00042.1]

Ibn-e-Baitar, (2003). Aljamaiul Mufradat-ul Advia Wal Aghzia. V.I (Urdu Trans), New Delhi, CCRUM; 248-252.

Igwenyi, I.O., Offor, C.E., Ajah, D.A. (2011). Chemical compositions of *Ipomoea aquatica* (green kangkong). *Int J Pharm Bio Sci, 2*(4), 593-598.

Imami, A., Taleb, A., Khalili, H. (2010). *PDR for Herbal Medicines.* (1st ed.). Tehran: Andisheh-Avar.

Iranshahy, M., Javadi, B., Iranshahi, M., Jahanbakhsh, S.P., Mahyari, S., Hassani, F.V., Karimi, G. (2017). A review of traditional uses, phytochemistry and pharmacology of *Portulaca oleracea* L. *J. Ethnopharmacol., 205*, 158-172.
[http://dx.doi.org/10.1016/j.jep.2017.05.004] [PMID: 28495602]

Irimpan, M.T., Jolly, C.I., Sheela, D. (2011). A study of the phytochemical composition and antibacterial activity of *Holostemma adakodien* Schultes. *Int. J. Pharm. Tech. Res., 3*(2), 1208-1210.

Isaac, L., Wood, C.W., Shannon, D.A. (2003). Pruning management effects on soil carbon and nitrogen in contour-hedgerow cropping with *Leucaena leucocephala* (Lam.) De Wit on sloping land in Haiti. *Nutr. Cycl. Agroecosyst., 65*(3), 253-263.
[http://dx.doi.org/10.1023/A:1022600720226]

Ishrat, J.B., Laizuman, N., Farhana, A.R., Obaydul, H. (2011). Antibacterial, Cytotoxic an Antioxidant Activity of Chloroform, n-hexane and Ethyl Acetate extract of plant *Amaranthus spinosus. Int. J. Pharm.*

Tech. Res., 3(3), 1675-1680.

Islam, M. (2019). Food and medicinal values of Roselle (*Hibiscus sabdariffa* L. Linne Malvaceae) plant parts: A review. *Open J. Nutr. Food Sci., 1*, 14-20.

Islam, M.T., Quispe, C., El-Kersh, D.M., Shill, M.C., Bhardwaj, K., Bhardwaj, P., Sharifi-Rad, J., Martorell, M., Hossain, R., Al-Harrasi, A., Al-Rawahi, A., Butnariu, M., Rotariu, L.S., Suleria, H.A.R., Taheri, Y., Docea, A.O., Calina, D., Cho, W.C. (2021). A literature-based update on *Benincasa hispida* (Thunb.) Cogn.: Traditional uses, nutraceutical, and phytopharmacological profiles. *Oxid. Med. Cell. Longev., 2021*(1), 6349041.
[http://dx.doi.org/10.1155/2021/6349041] [PMID: 34925698]

Islam, MT, Riaz, TA, Ayatollahi, SA, Sharifi-Rad, J (2021). Anxiolytic-like effect of *Urena lobata* (L.) in swiss albino mice. *Clin. phytosci, 7*(1), 1-6.

Islam, M., Uddin, M. (2017). A revision on *Urena lobata* L. *Int. J. Med. (Dubai), 5*(1), 126-131.
[http://dx.doi.org/10.14419/ijm.v5i1.7525]

Iwashina, T., Konishi, T., Takayama, A., Fukada, M., Ootani, S. (1999). Isolation and identification of the flavanoids in the leaves of taro. *Ann. Tsukuba Bot. Gard., 18*, 71-74.

Izquierdo-Vega, J., Arteaga-Badillo, D., Sánchez-Gutiérrez, M., Morales-González, J., Vargas-Mendoza, N., Gómez-Aldapa, C., Castro-Rosas, J., Delgado-Olivares, L., Madrigal-Bujaidar, E., Madrigal-Santillán, E. (2020). Organic Acids from Roselle (*Hibiscus sabdariffa* L.)—A Brief Review of Its Pharmacological Effects. *Biomedicines, 8*(5), 100.
[http://dx.doi.org/10.3390/biomedicines8050100] [PMID: 32354172]

Jadhao, K., Wadekar, M., Mahalkar, M. (2008). Comparative study of availability of vitamins from *Leea macrophylla* Roxb. *Biosci. Biotechnol. Res. Asia, 6*, 847-849.

Jain, S.C., Jain, R., Mascolo, N., Capasso, F., Vijayvergia, R., Sharma, R.A., Mittal, C. (1998). Ethnopharmacological evaluation of *Pergularia daemia* (Forsk.) Chiov. *Phytother. Res., 12*(5), 378-380.
[http://dx.doi.org/10.1002/(SICI)1099-1573(199808)12:5<378::AID-PTR308>3.0.CO;2-Z]

Joshi, G.P., Rawat, M.S., Bisht, V.K., Negi, J.S., Singh, P. (2010). Chemical constituents of *Asparagus*. *Pharmacogn. Rev., 4*(8), 215-220.
[http://dx.doi.org/10.4103/0973-7847.70921] [PMID: 22228964]

Jadhav, V.B., Thakare, V.N., Suralkar, A.A., Deshpande, A.D., Naik, S.R. (2010). Hepatoprotective activity of *Luffa acutangula* against CCl4 and rifampicin induced liver toxicity in rats: a biochemical and histopathological evaluation. *Indian J. Exp. Biol., 48*(8), 822-829.
[PMID: 21341541]

Jadhav, S., Chavan, N. (2013). Evaluation of antimicrobial activity of *Luffa acutangula* (L.) Roxb. var. amara (Roxb.). *Clarke. Int. J. Adv. Res., 1*, 323-326.

Jagatha, G., Senthilkumar, N. (2011). Evalution of anti-diabetic activity of methanol extract of *Digera muricata* (l) Mart in alloxan Induced diabetic rats. *Int. J. Pharma Sci., 2*(6), 1525-1529.
[http://dx.doi.org/10.13040/IJPSR.0975-8232.2(6).1525-29]

Jagtap, S.D., Deokule, S.S., Bhosle, S.V. (2008). Ethnobotanical uses of endemic and RET plants by the Korku tribe of Amravati district, Maharashtra. *Indian J. Tradit. Knowl., 7*, 284-287.

Jain, P.S., Bari, S.B. (2009). Antibacterial and antifungal activity of extracts of woody stem of *Wrightia tinctoria* R. Br. *Int J. Pharm Recent Res., 1*, 18-21.

Jain, R., Sharma, A., Gupta, S., Sarethy, I.P., Gabrani, R. (2011). *Solanum nigrum*: current perspectives on therapeutic properties. *Altern. Med. Rev., 16*(1), 78-85.
[PMID: 21438649]

Jaiswal, V. (2010). Culture and ethnobotany of Jaintia tribal community of Meghalaya, Northeast India-A mini review. *Indian J. Tradit. Knowl., 9*(1), 38-44.

Jamadagni, P., Pawar, S., Jamadagni, S., Chougule, S., Gaidhani, S., Murthy, S.N. (2017). Review of *Holarrhena antidysenterica* (L.) Wall. ex A. DC.: Pharmacognostic, pharmacological, and toxicological perspective. *Pharmacogn. Rev., 11*(22), 141-144.
[http://dx.doi.org/10.4103/phrev.phrev_31_16] [PMID: 28989249]

Jamil, S.S., Nizami, Q., Salam, M. (2007). *Centella asiatica* (Linn.) Urban: a review. *Nat. Prod. Radiance, 6*(2), 158-170.

Jamkhande, P.G., Barde, S.R., Patwekar, S.L., Tidke, P.S. (2013). Plant profile, phytochemistry and pharmacology of *Cordia dichotoma* (Indian cherry): A review. *Asian Pac. J. Trop. Biomed., 3*(12), 1009-1012.
[http://dx.doi.org/10.1016/S2221-1691(13)60194-X] [PMID: 24093795]

Janarthanan, L., Venkateswarlu, B.S. (2018). Pharmacognostical, physicochemical investigations and phytochemical of seeds of *Meyna laxiflora* Robyns. *Pharmacol. Pharmacother., 7*(4), 455-465.

Jayasree, T., Kishore, K.K., Vinay, M., Vasavi, P., Chandrasekhar, N., Manohar, V.S., Dixit, R. (2011). Evaluation of the Diuretic effect of the chloroform extract of the *Benincasa hispida* rind (Pericarp) Extract in Guinea-pigs. *J. Clin. Diagn. Res., 5*(3), 578-582.

Jayvir, A., Minoo, P., Gauri, B., Ripal, K. (2007). *Natural Heals: A glossary of selected indigenous medicinal plant of India.* (2nd ed., p. 22). Ahmadabad, India: SRIST Innovations.

Jeevanantham, P., Vincent, S., Balasubramaniam, A., Jayalakshmi, B., Kumar, N.S. (2011). Anti-cancer activity of methanolic extract of aerial parts of *Momordica cymbalaria* Hook F. against Ehrlich ascites carcinoma in mice. *Int. J. Pharma Sci., 3*(8), 1408.

Jeyaprakash, K., Ayyanar, M., Geetha, K.N., Sekar, T. (2011). Traditional uses of medicinal plants among the tribal people in Theni District (Western Ghats), Southern India. *Asian Pac. J. Trop. Biomed., 1*(1), S20-S25.
[http://dx.doi.org/10.1016/S2221-1691(11)60115-9]

Jeyaweera DMA. (1981). Sri Lanka: National Science Council of Sri Lanka; Medicinal plants used in Ceylon Colombo.

Jha, D.K., Koneri, R., Samaddar, S. (2018). Medicinal use of an ancient herb *Momordica Cymbalaria*: a review. *Int. J. Pharm. Sci. Res., 9*(2), 432-441.

Jha DK, Koneri R, Samaddar S. (2017). Potential Bio-Resources of *Momordica dioica* Roxb: A Review. *Int J Pharm Sci Rev Res., 45*(37): 203-209.

Jiang, J., Xu, Q. (2003). Immunomodulatory activity of the aqueous extract from rhizome of *Smilax glabra* in the later phase of adjuvant-induced arthritis in rats. *J. Ethnopharmacol., 85*(1), 53-59.
[http://dx.doi.org/10.1016/S0378-8741(02)00340-9] [PMID: 12576202]

Jivani, N.P. (2011). Phytopharmacological properties of *Bambusa arundinacea* as a potential medicinal tree: An overview. *J. Appl. Pharm. Sci., 2011*, 27-31.

Jose, J.K., Kuttan, R. (2000). Hepatoprotective activity of *Emblica officinalis* and Chyavanaprash. *J. Ethnopharmacol., 72*(1-2), 135-140.
[http://dx.doi.org/10.1016/S0378-8741(00)00219-1] [PMID: 10967464]

Joshi, S.G. (1997). Medicinal plants. New Delhi. *Oxford & IBH Publication., 3,* 126-127.

Joshi, S.G. (2007). *Medicinal Plants.* (p. 243). New Delhi: Oxford & IBH Publishing Co. Pvt. Ltd.

Jun, M.S., Ha, Y.M., Kim, H.S., Jang, H.J., Kim, Y.M., Lee, Y.S., Kim, H.J., Seo, H.G., Lee, J.H., Lee, S.H., Chang, K.C. (2011). Anti-inflammatory action of methanol extract of *Carthamus tinctorius* involves in heme oxygenase-1 induction. *J. Ethnopharmacol., 133*(2), 524-530.
[http://dx.doi.org/10.1016/j.jep.2010.10.029] [PMID: 20969944]

Junaid, N., Singh, P., Bansal, Y., Goel, R.K. (2009). Anti-inflammatory, analgesic and antipyretic activity of aqueous extracts of fresh leaves of *Coccinia indica*. *Inflammopharmacology, 17*, 219-244.
[http://dx.doi.org/10.1007/s10787-009-0010-3]

Kabiruddin, M. (1951). *Makhzanul Mufradat.* (p. 266). Lahore: Sheikh Mohammad Bashir & Sons.

Kaewamatawong, R., Kitajima, M., Kogure, N., Takayama, H. (2008). Flavonols from *Bauhinia malabarica*. *J. Nat. Med., 62*(3), 364-365.
[http://dx.doi.org/10.1007/s11418-008-0249-9] [PMID: 18414979]

Kala, A., Prakash, J. (2004). Nutrient composition and sensory profile of differently cooked green leafy vegetables. *Int. J. Food Prop., 7*(3), 659-669.
[http://dx.doi.org/10.1081/JFP-200033079]

Kalariya, M., Parmar, S., Sheth, N. (2010). Neuropharmacological activity of hydroalcoholic extract of leaves of *Colocasia esculenta*. *Pharm. Biol., 48*(11), 1207-1212.
[http://dx.doi.org/10.3109/13880201003586887] [PMID: 20818937]

Kale, N., Rathod, S., More, S., Shinde, N. (2021). Phyto-Pharmacological Profile of *Wrightia tinctoria*. *J Res Pharm Sci., 11*(4), 301-308.
[http://dx.doi.org/10.52711/2231-5659.2021.00047]

Kalle, R., Belichenko, O., Kuznetsova, N., Kolosova, V., Prakofjewa, J., Stryamets, N., Mattalia, G., Šarka, P., Simanova, A., Prūse, B., Mezaka, I., Sõukand, R. (2020). Gaining momentum: Popularization of *Epilobium angustifolium* as food and recreational tea on the Eastern edge of Europe. *Appetite, 150*, 104638.
[http://dx.doi.org/10.1016/j.appet.2020.104638] [PMID: 32113918]

Kamal, S., Akhter, R., Tithi, N.A. (2015). Biological investigations of the leaf extract of *Spondias pinnata*. *IJPSR, 6*(8), 3351-3358.
[http://dx.doi.org/10.13040/IJPSR.0975-8232.6(8).3351-58]

Kamaruddin, H.S., Megawati, M., Nurliana, N., Sabandar, C.W. (2021). Chemical Constituents and Antioxidant Activity of *Melothria scabra* Naudin Fruits. *Borneo J. Phar., 4*(4), 283-292.
[http://dx.doi.org/10.33084/bjop.v4i4.2890]

Kamble, A., Lobo, V. (2022). *Smilax zeylanica* linn: pharmacognostic, taxonomical and phytochemical status. research & reviews. *J Herb Sci., 11*(1), 1-4p.

Kamble, S.Y., More, T.N., Patil, S.R., Pawar, S.G., Bindurani, R., Bodhankar, S.L. (2009). Plants used by the tribes of Northwest Maharashtra for the treatment of gastrointestinal disorders. *Indian J. Tradit. Knowl., 7*(2), 321-325.

Kamel, W., Abd El-Ghani, M.M., El-Bous, M. (2009). Taxonomic study of Capparaceae from Egypt: Revisited. *African J. Pl. Sci. Biotech., 3*, 27-35.

Kanakhara, R.D., Harisha, C.R., Shukla, V.J., Acharya, R.N. (2018). A complete pharmacognostical profile of *Telosma pallida* (L.) Kurz. (Stem)-An extrapharmacopoeial drug. *Int. J. Res. Ayurveda Pharm., 9*(1), 36-43.
[http://dx.doi.org/10.7897/2277-4343.0918]

Kanakhara, R., Rudrappa, H., Shukla, V., Acharya, R. (2018). Detailed pharmacognostical and analytical profile of *telosma pallida* (l.) kurz. (leaf): A folklore medicinal plant of Gujarat State. *Anc. Sci. Life, 37*(3), 120-126.
[http://dx.doi.org/10.4103/asl.ASL_97_17]

Kapoor, L.D. (1990). *Handbook of Ayurvedic Medicinal plants..* London: CRC Press.

Karimi, E., Jaafar, H.Z.E., Ahmad, S. (2013). Antifungal, anti-inflammatory and cytotoxicity activities of three varieties of *labisia pumila* benth: from microwave obtained extracts. *BMC Complement. Altern. Med., 13*(1), 20.
[http://dx.doi.org/10.1186/1472-6882-13-20] [PMID: 23347830]

Karkal, Y.R., Bairy, L.K. (2007). Safety of aqueous extract of *Tinospora cordifolia* (Tc) in healthy volunteers: A double-blind randomised placebo-controlled study. *Iranian J Pharmacol Therap., 6*, 59-61.

Karou, S.D., Tchacondo, T., Ilboudo, D.P., Simpore, J. (2011). Sub-Saharan Rubiaceae: a review of their traditional uses, phytochemistry and biological activities. *Pak. J. Biol. Sci., 14*(3), 149-169.
[http://dx.doi.org/10.3923/pjbs.2011.149.169] [PMID: 21870639]

Kasahara, Y., Kumaki, K., Sato, T., Katagiri, S. (1989). Pharmacological studies on flower petals of *Carthamus tinctorius* central actions and anti-inflammation. *Shoyakugaku Zasshi., 43*, 331-338.

Kassuya, C.A., Leite, D.F., de Melo, L.V., Rehder, V.L., Calixto, J.B. (2005). Anti-inflammatory properties of extracts, fractions and lignans isolated from *Phyllanthus amarus. Planta Med., 71*(8), 721-726.
[http://dx.doi.org/10.1055/s-2005-871258] [PMID: 16142635]

Kathiresan, K., Ravishankar, G.A., Venkataraman, L.V. (1997). In vitro multiplication of a coastal plant Sesuvium portulacastrum L. by axillary buds.*Biotechnological applications of plant tissue and cell culture.* (pp. 185-192). India: Oxford and IBH Publishing C., Pvt. Ltd..

Kavalan, R., Jaleel, V.A., Gothandam, K.M. (2018). Morphological, phytochemical and antibacterial properties of *Amorphophallus sylvaticus* (araceae). *Int. J. Adv. Res. (Indore), 6*(10), 886-894.
[http://dx.doi.org/10.21474/IJAR01/7884]

Kavitha, D., Shilpa, P.N., Devaraj, S.N. (2004). Antibacterial and antidiarrhoeal effects of alkaloids of *Holarrhena antidysenterica* WALL. *Indian J. Exp. Biol., 42*(6), 589-594.
[PMID: 15260110]

Kekuda, P.T.R., Raghavendra, H.L., Bharadwaj, N.A., Akhilesha, S. (2018). Traditional uses, chemistry and pharmacological activities of *Leea indica* (Burm. f.) Merr. (Vitaceae): A comprehensive review. *Int. J. Green Pharm., 12*(01).
[http://dx.doi.org/10.22377/ijgp.v12i01.1602]

Kemboi, D., Peter, X., Langat, M., Tembu, J. (2020). A review of the ethnomedicinal uses, biological activities, and triterpenoids of Euphorbia species. *Molecules, 25*(17), 4019.

[http://dx.doi.org/10.3390/molecules25174019] [PMID: 32899130]

Kemisetti, D., Rajeswar Das, D., Bolay, B. (2022). A comprehensive review on *Musa paradisiaca* taxonomical, morphological classification and its pharmacological activities. *J. Pharm. Negat. Results, 13*(S10), 737-749.
[http://dx.doi.org/10.47750/pnr.2022.13.S10.079]

Kengoh, J.B., Peter Etape, E., Victorine Namondo, B., Foba-Tendo, J., Nafu, Y.R., Fabien, B.E. (2021). Influence of *Urena lobata* fibre treatment on mechanical performance development in Hybrid *Urena lobata*: fibre/gypsum plaster composites. *Adv. Mater. Sci. Eng., 2021*(1), 5514525.
[http://dx.doi.org/10.1155/2021/5514525]

Kenjale, R., Shah, R., Sathaye, S. (2008). Effects of *Chlorophytum borivilianum* on sexual behaviour and sperm count in male rats. *Phytother. Res., 22*(6), 796-801.
[http://dx.doi.org/10.1002/ptr.2369] [PMID: 18412148]

Kenjale, R.D., Shah, R.K., Sathaye, S.S. (2007). Anti-stress and anti-oxidant effects of roots of *Chlorophytum borivilianum* (Santa Pau & Fernandes). *Indian J. Exp. Biol., 45*(11), 974-979.
[PMID: 18072542]

Khan, I.A., Abourashed, E.A. (2010). *Leung's Encyclopedia of Common Natural Ingredients: Used in Food, Drugs and Cosmetics.* (3rd ed., pp. 52-53). Hoboken, NJ, USA: Wiley.

Khan, M.N., Ali, S., Yaseen, T., Ullah, S., Zaman, A., Iqbal, M., Shah, S. (2020). Eco-taxonomic study of family Poaceae (Gramineae). *RADS Journal of Biological Research & Applied Sciences, 10*(2), 63-75.
[http://dx.doi.org/10.37962/jbas.v10i2.191]

Khan, M.S., Alam, M.K. (1996). *Homestead Flora of Bangladesh. Publ. BARC, IDRC.* (p. 144). Dhaka: SDC.

Khan, N.S., Chaurasia, B., Dixit, A.K. (2021). Pharmacognostic Characterization for Taxonomic Identification of *Bacopa monnieri* (L.) Wettst. for Quality Control. (2021). *Int. J. Life Sci. Pharma Res., 11*(1), L54-L62.
[http://dx.doi.org/10.22376/ijpbs/lpr.2021.11.1.L54-62]

Khanam, Z., Singh, O., Singh, R., Bhat, I.U.H. (2013). Safed musli (*Chlorophytum borivilianum*): A review of its botany, ethnopharmacology and phytochemistry. *J. Ethnopharmacol., 150*(2), 421-441.
[http://dx.doi.org/10.1016/j.jep.2013.08.064] [PMID: 24045177]

Khare, C.P. (2004). *Indian herbal remedies: rational Western therapy, ayurvedic, and other traditional usage, Botany..* Springer science & business media.
[http://dx.doi.org/10.1007/978-3-642-18659-2]

Khare, C.P. (2004). *Encyclopedia of Indian medicinal plants: rational western therapy, ayurvedic and other traditional usage, botany..* Springer.

Khare, C.P. (2006). *Indian Medicinal Plants, an Illustrated Dictionary.*

Khare CP. (2007). Indian Medicinal Plants. An Illustrated Dictionary Springer, New York; pp. 139-140.
[http://dx.doi.org/10.1007/978-0-387-70638-2]

Khare, E., Ghosh, S., Sharma, A., K, N. (2020). The versatility of *Tricholepis glaberrima* (Brahmadandi): An Overview. *Environ. Conserv. J., 21*(3), 149-154.
[http://dx.doi.org/10.36953/ECJ.2020.21318]

Khatun, M.T., Siddiqi, M.M.A., Ma, A-M., Sohrab, M.H., Rahman, A.F.M.M., Hasan, C.M., Chowdhury, A.M.S. (2013). New diarylheptanoid from *Garuga pinnata* roxb. *Dhaka University Journal of Science, 61*(2), 131-134.
[http://dx.doi.org/10.3329/dujs.v61i2.17058]

Khond, M., Bhosale, J.D., Arif, T., Mandal, T.K., Padhi, M.M., Dabur, R. (2009). Screening of some selected medicinal plants extracts for *in-vitro* antimicrobial activity. *Middle East J. Sci. Res., 4*(4), 271-278.

Khyade MS. (2006). Pharmacognostic studies of some plants of Aurangabd district-II [dissertation]. Aurangabad: Dr. Babasaheb Ambedkar Marathwada University.

Kim, K.H., Moon, E.J., Kim, S.Y., Lee, K.R. (2010). Anti-melanogenic fatty acid derivatives from the tuber-barks of *Colocasia antiquorum* var. esculenta. *Bull. Korean Chem. Soc., 31*(7), 2051-2053.
[http://dx.doi.org/10.5012/bkcs.2010.31.7.2051]

Kirthikar, K.R., Basu, B.D., II (1969). 2nd ed. New Delhi: Periodical Experts, Jayyed Press. Indian Medicinal Plants; 1 p. 623.

Kirtikar, K.R., Basu, B.D. (2006). (2nd ed.), Indian medicinal plants, Lalit Mohan Basu, Allahabad; 3 pp. 2201-2204.

Kirtikar, K.R., Basu, B.D. (1998). Data on medicinal plants and chemical constituents. *Indian medicinal plants.* VIMSAT Publishers.

Kirtikar, K.R., Basu, B.D. (1993). Indian Medicinal Plants. 2nd Ed. Deheradun: International Publisher; 1 pp. 200-201.

Kirtikar, KR, Basu, BD (2003). Dehradun, India: Oriental Enterprises. Indian medicinal plants with illustrations.

Kirtikar, K.R., Basu, B.D. (1987). Indian Medical plants, Dehradun. *International Book Publication Distribution, 1*, 195-201.

Kirtikar, K.R., Basu, B.D. (1975). *Indian Medicinal Plants.* (2nd ed., pp. 2327-2328). Dehra Dun: Bishen Singh & Mahendra Pal Singh.

Kirtikar, K.R., Basu, B.D. (1988). *Indian Medicinal Plants., II*, 1162-1163.

Kirtikar, K.R., Basu, B.D. (1987). *Indian Medicinal plants.*

Kirtikar, K.R., Basu, B.D. (1981). *Indian medicinal plants.* Periodical Experts Book Agency.

Kirtikar, K.R., Basu, B.D. (1975). Indian Medicinal Plants. *Allabahad: Lalit Mohen Basu, 2*, 1180-1181.

Kirtikar, K.R., Basu, B.D. (1975). Indian medicinal plants. *Bishen Mahendra Pal Singh, Dehradun, 2*, 842-844.

Kirtikar, K.R., Basu, B.D. (1994). *Indian Medicinal Plants.* (pp. 1908-1909). Dehradun: International Book Distributors.

Kishore, H., Kaur, B., Kalsi, V., Suttee, A. (2016). Effect of Ethanolic Extract of *Spondias pinnata* on Ischemia-Reperfusion Injury and Ischemic Preconditioning of Heart. *Inter J. Phcog Phyto Res., 8*(5), 865-870.

Kokate, C.K., Purohit, A.P., Gokhale, S.B. (2006). *Text book of Pharmacognosy*. (26th ed., pp. 593-597). Pune: Nirali Prakashan.

Korkina, L.G., Pastore, S., Dellambra, E., De Luca, C. (2013). New molecular and cellular targets for chemoprevention and treatment of skin tumors by plant polyphenols: a critical review. *Curr. Med. Chem., 20*(7), 852-868.
[PMID: 23210776]

Kotiya A, Solanki Y, Reddy GV. (2020). Flora of Rajasthan. Rajasthan state biodiversity board: 1-769.

Kowalczyk, E., Kopff, A., Fijałkowski, P., Kopff, M., Niedworok, J., Błaszczyk, J., Kedziora, J., Tyślerowicz, P. (2003). Effect of anthocyanins on selected biochemical parameters in rats exposed to cadmium. *Acta Biochim. Pol., 50*(2), 543-548.
[http://dx.doi.org/10.18388/abp.2003_3707] [PMID: 12833179]

Krishna, A.R., Singh, A., Jaleel, V.A., Raj, S., Karthikeyan, S., Gothandam, K.M. (2013). Morphological, phytochemical, and anti-bacterial properties of wild and indigenous plant (*Amorphophallus commutatus*). *J. Med. Plants Res., 7*, 744-748.

Krishna, C.B. (2012). Anti-inflammatory activity of *Basella alba* L. in albino rats. *J. Appl. Pharm. Sci., 2*, 87-89.
[http://dx.doi.org/10.7324/JAPS.2012.2413]

Krishna, K.L., Paridhavi, M., Patel, J.A. (2008). Review on nutritional, medicinal and pharmacological properties of papaya (*Carica papaya* Linn.). *Nat. Prod. Radiance, 7*(4), 364-373.

Krishnaveni, A., Santh, R.T. (2009). Pharmacognostical and Preliminary Phytochemical Studies of *Argyreia nervosa* Burm, Ethanobot. *Leaflets, 13*, 293.

Kritchevsky, D. (1978). Fiber, lipids, and atherosclerosis. *Am. J. Clin. Nutr., 31*(10) (Suppl.), S65-S74.
[http://dx.doi.org/10.1093/ajcn/31.10.S65] [PMID: 101075]

Kruawan, K., Kangsadalampai, K. (2006). Antioxidant activity, phenolic compound contents and antimutagenic activity of some water extract of herbs. *Thaiphesatchasan, 30*, 28-35.

Kubitzki, K. (1993). Plumbaginaceae.*Flowering Plants· Dicotyledons: Magnoliid, Hamamelid and Caryophyllid Families.* (pp. 523-530). Berlin, Heidelberg: Springer Berlin Heidelberg.
[http://dx.doi.org/10.1007/978-3-662-02899-5_62]

Kubitzki, K. (1998). Typhaceae.*Flowering Plants· Monocotyledons: Alismatanae and Commelinanae (except Gramineae).* (pp. 457-461). Berlin, Heidelberg: Springer Berlin Heidelberg.
[http://dx.doi.org/10.1007/978-3-662-03531-3_46]

Kulkarni, R.S., Gangaprasad, S., Swamy, G.S. (1993). *Tamarindus indica*: Economically an important minor forest product. *Minor Forest Prod News., 3*, 6.

Kulkarni, V.M., Rathod, V.K. (2014). Extraction of mangiferin from Mangifera indica leaves using three phase partitioning coupled with ultrasound. *Ind. Crops Prod., 52*, 292-297.
[http://dx.doi.org/10.1016/j.indcrop.2013.10.032]

Kumar, A., Saluja, A.K., Shah, U.D., Mayavanshi, A.V. (2007). Pharmacological potential of *Albizzia lebbeck*: a review. *Phacog Rev., 1*(1), 171-174.

Kumar, B., Yadav, D.K., Govindarajan, R., Pushpangadan, P. (2006). Wound healing activity of *Pergularia daemia* (Forsk). *Chiov. Pharmacog. Ethnopharmacol., 1*, 12-14.

Kumar, B.M., Kumar, S.S., Fisher, R.F. (1998). Intercropping teak with *Leucaena* increases tree growth and modifies soil characteristics. *Agrofor. Syst., 42*(1), 81-89.
[http://dx.doi.org/10.1023/A:1006199910985]

Kumar, G.P., Shiddamallayya, N. (2016). Survey of wild edible fruits in Hassan Forest division, Karnataka, India. *J. Biodivers. Environ. Sci., 8*, 57-66.

Kumar, K.G., Boopathi, T. (2018). An updated overview on pharmacognostical and pharmacological screening of *Tecoma* stans. *Pharmatutor, 6*(1), 38-49.
[http://dx.doi.org/10.29161/PT.v6.i1.2018.38]

Kumar, N., Rungseevijitprapa, W., Narkkhong, N.A., Suttajit, M., Chaiyasut, C. (2012). 5α-reductase inhibition and hair growth promotion of some Thai plants traditionally used for hair treatment. *J. Ethnopharmacol., 139*(3), 765-771.
[http://dx.doi.org/10.1016/j.jep.2011.12.010] [PMID: 22178180]

Kumar, P. (2014). Reddy. Y N. Protective effect of *Canavalia gladiata* (sword bean) fruit extracts and its flavanoidal contents, against azathioprine-induced toxicity in hepatocytes of albino rats. *Toxicol. Environ. Chem., 96*(3), 474-481.
[http://dx.doi.org/10.1080/02772248.2014.950805]

Kumar, R., Kumar, S., Patra, A., Jayalakshmi, S. (2009). Hepatoprotective activity of aerial parts of *Plumbago zeylanica* linn against carbon tetrachlorideinduced hepatotoxicity in rats. Int. *J. Pharm. Pharm. Sci., 1*, 171-175.

Kumar, S., Behera, S.P., Jena, P.K. (2013). Validation of tribal claims on *Dioscorea pentaphylla* L. through phytochemical screening and evaluation of antibacterial activity. *Plant Sci Res., 35*(1&2), 55-61.

Kumar, S., Das, G., Shin, H.S., Patra, J.K. (2017). *Dioscorea* spp. (a wild edible tuber): a study on its ethnopharmacological potential and traditional use by the local people of Similipal Biosphere Reserve, India. *Front. Pharmacol., 8*, 52.
[http://dx.doi.org/10.3389/fphar.2017.00052] [PMID: 28261094]

Kumar, S., Jena, P.K. (2017). Tools from Biodiversity: Wild Nutraceutical Plants. Ed: James N Furze et al. Identifying Frontier Research Integrating Mathematic Approaches to Diverse Systems / Sustainability. Springer, Switzerland.
[http://dx.doi.org/10.1007/978-3-319-43901-3-9]

Kumar, S., Malhotra, R., Kumar, D. (2010). *Euphorbia hirta*: Its chemistry, traditional and medicinal uses, and pharmacological activities. *Pharmacogn. Rev., 4*(7), 58-61.
[http://dx.doi.org/10.4103/0973-7847.65327] [PMID: 22228942]

Kumar, S., Tripathy, P.K., Jena, P.K. (2012). Study of wild edible plants among tribal groups of Simlipal Biosphere Reserve Forest, Odisha, India; with special reference to *Dioscorea* species. *Int J Biol Tech., 3*(1), 11-19.

Kumar, S. (2016). Yam (*Dioscorea* Species): Future Functional Wild Food of Tribal Odisha, India. *Front Bioactive Comp., 2*, 186-208.

Kumar, S.S., Manoj, P., Giridhar, P. (2015). Nutrition facts and functional attributes of foliage of *Basella* spp. *Lebensm. Wiss. Technol., 64*(1), 468-474.
[http://dx.doi.org/10.1016/j.lwt.2015.05.017]

Kumaran, A., Karunakaran, R.J. (2006). Nitric oxide radical scavenging active components from *Phyllanthus emblica* L. *Plant Foods Hum. Nutr., 61*(1), 1-5.
[http://dx.doi.org/10.1007/s11130-006-0001-0] [PMID: 16688481]

Kushwaha, S.K., Jain, A., Gupta, V.B., Patel, J.R. (2005). Hepatoprotective activity of the fruits of *Momordica dioica. Niger. J. Nat. Prod. Med., 9*, 29-31.
[http://dx.doi.org/10.4314/njnpm.v9i1.11830]

Lampariello, L.R., Cortelazzo, A., Guerranti, R., Sticozzi, C., Valacchi, G. (2012). The magic velvet bean of *Mucuna pruriens. J. Tradit. Complement. Med., 2*(4), 331-339.
[http://dx.doi.org/10.1016/S2225-4110(16)30119-5] [PMID: 24716148]

Lanhers, M.C., Fleurentin, J., Dorfman, P., Mortier, F., Pelt, J.M. (1991). Analgesic, antipyretic and anti-inflammatory properties of *Euphorbia hirta. Planta Med., 57*(3), 225-231.
[http://dx.doi.org/10.1055/s-2006-960079] [PMID: 1896520]

Larsen, K., Lock, J.M., Maas, H., Maas, P.J. (1998). Zingiberaceae.*Flowering Plants· Monocotyledons: Alismatanae and Commelinanae (except Gramineae).* (pp. 474-495). Berlin, Heidelberg: Springer Berlin Heidelberg.
[http://dx.doi.org/10.1007/978-3-662-03531-3_49]

Lather, A., Gupta, V., Bansal, P., Singh, R., Chaudhary, A.K. (2010). Pharmacological potential of ayurvedic formulation: Kutajghan vati-A review. *Int. J. Adv. Sci. Res., 1*(02), 41-45.

Lawrence, R.M., Chaudhary, S. (2004). *Caralluma fimbriata* in the treatment of obesity 12th Annu WConger on Antiaging medicine held on December 2-5, (Las vegas, NV USA).

Lee, A.S., Kim, J.S., Lee, Y.J., Kang, D.G., Lee, H.S. (2012). Anti-TNF-α activity of *Portulaca oleracea* in vascular endothelial cells. *Int. J. Mol. Sci., 13*(5), 5628-5644.
[http://dx.doi.org/10.3390/ijms13055628] [PMID: 22754320]

Lee, J.Y., Chang, E.J., Kim, H.J., Park, J.H., Choi, S.W. (2002). Antioxidative flavonoids from leaves of *Carthamus tinctorius. Arch. Pharm. Res., 25*(3), 313-319.
[http://dx.doi.org/10.1007/BF02976632] [PMID: 12135103]

Lee, S., Wee, W., Yong, J., Syamsumir, D. (2011). Antimicrobial, antioxidant, anticancer property and chemical composition of different parts (corm, stem and leave) of *Colocasia esculenta* extract. *Ann Univ Mariae Curie-Sklodowska Pharm., 24*(3), 9-16.

Leporatti, M.L., Ghedira, K. (2009). Comparative analysis of medicinal plants used in traditional medicine in Italy and Tunisia. *J. Ethnobiol. Ethnomed., 5*(1), 31.
[http://dx.doi.org/10.1186/1746-4269-5-31] [PMID: 19857257]

Leu, Y.L., Hwang, T.L., Kuo, P.C., Liou, K.P., Huang, B.S., Chen, G.F. (2012). Constituents from *Vigna vexillata* and their anti-inflammatory activity. *Int. J. Mol. Sci., 13*(8), 9754-9768.
[http://dx.doi.org/10.3390/ijms13089754] [PMID: 22949828]

Li, A., Wei, P., Hsu, H.C., Cooks, R.G. (2013). Direct analysis of 4-methylimidazole in foods using paper spray mass spectrometry. *Analyst (Lond.), 138*(16), 4624-4630.
[http://dx.doi.org/10.1039/c3an00888f] [PMID: 23762894]

Li, H., Hwang, S., Kang, B., Hong, J., Lim, S. (2014). Inhibitory effects of *Colocasia esculenta* (L.) Schott constituents on aldose reductase. *Molecules, 19*(9), 13212-13224.

[http://dx.doi.org/10.3390/molecules190913212] [PMID: 25255750]

Liang, N., Yang, X.X., Wang, G.C., Wu, X., Yang, Y.T., Luo, H.J., Li, Y.L. (2012). [Study on the chemical constituents of *Elephantopus mollis*.]. *Zhong Yao Cai, 35*(11), 1775-1778. [PMID: 23627086]

Lima, C.C., Lemos, R.P. (2014). Conser Liang va LM. Dilleniaceae family: an overview of its ethnomedicinal uses, biological and phytochemical profile. *J Phacog Phytochem., 3*(2), 181-204.

Lipińska, M., Wibowo, A.R., Margońska, H. (2022). Notes on the genus *Nervilia* (Orchidaceae, Nervilieae) in Bali with new records. *Acta Soc. Bot. Pol., 91.* [http://dx.doi.org/10.5586/asbp.915]

Lipkin, A., Anisimova, V., Nikonorova, A., Babakov, A., Krause, E., Bienert, M., Grishin, E., Egorov, T. (2005). An antimicrobial peptide Ar-AMP from amaranth (*Amaranthus retroflexus* L.) seeds. *Phytochemistry, 66*(20), 2426-2431. [http://dx.doi.org/10.1016/j.phytochem.2005.07.015] [PMID: 16126239]

Little, E.L., Wadsworth, F.W. (1964). Washington DC: US Department of Agriculture. Common Trees of Puerto Rico and the Virgin Islands, Agriculture Handbook.

Litvinenko, Y.A., Muzychkina, R.A. (2003). MuzychKina RA. Phytochemical investigation of biologically active substances in certain Kazakhstan Rumex species. 1. *Chem. Nat. Compd., 39*(5), 446-449. [http://dx.doi.org/10.1023/B:CONC.0000011117.01356.4c]

Lokhande, V.H., Gor, B.K., Desai, N.S., Nikam, T.D., Suprasanna, P. (2013). *Sesuvium portulacastrum*, a plant for drought, salt stress, sand fixation, food and phytoremediation. A review. *Agronomy (Basel), 33*, 329-348. [http://dx.doi.org/10.1007/s13593-012-0113-x]

Madhavan, V., Hemalatha, H.T., Gurudeva, M.R., Yoganarasimhan, S.N. (2010). Pharmacognostical studies on the rhizome and root of *Smilax zeylanica* Linn.–A potential alternate source for the Ayurvedic drug Chopachinee. *Indian J. Nat. Prod. Resour., 1*(3), 328-337.

Madhuri, S., Pandey, G., Verma, K.S. (2011). Antioxidant, immunomodulatory and anticancer activities of *Emblica officinalis*: an overview. *Int Res J Pharm., 2*(8), 38-42.

Mahato, D., Sharma, H.P. (2019). Phytochemical profiling and antioxidant activity of *Leea macrophylla* Roxb. ex Hornem. *in vitro* study. *Indian J. Tradit. Knowl., 18*(3), 493-499.

Mahesh, R., Muthuchelia, K., Maridass, M., Raju, G. (2010). *In vitro* Propagation of Wild Yam, *Dioscorea wightii* Through Nodal Cultures. *Int J Biol Tech, 1*(1), 111-113.

Maheshu, V., Priyadarsini, D.T., Sasikumar, J.M. (2014). Antioxidant capacity and amino acid analysis of *Caralluma adscendens* (Roxb.) Haw var. fimbriata (wall.) Grav. & Mayur. aerial parts. *J. Food Sci. Technol., 51*(10), 2415-2424. [http://dx.doi.org/10.1007/s13197-012-0761-5] [PMID: 25328180]

Mahishi, P., Srinivasa, B.H., Shivanna, M.B. (2005). Medicinal plant wealth of local communities in some villages in Shimoga District of Karnataka, India. *J. Ethnopharmacol., 98*(3), 307-312. [http://dx.doi.org/10.1016/j.jep.2005.01.035] [PMID: 15814264]

Maji, P., Ghosh Dhar, D., Misra, P., Dhar, P. (2020). *Costus speciosus* (Koen ex. Retz.) Sm.: Current status and future industrial prospects. *Ind. Crops Prod., 152*, 112571.

[http://dx.doi.org/10.1016/j.indcrop.2020.112571]

Makinde, E.A., Ayeni, L.S., Ojeniyi, S.O. (2010). Morphological characteristics of *Amaranthus cruentus* L. as influenced by kola pod husk, organomineral and NPK fertilizers in southwestern Nigeria. *N Y Sci J., 3*(5), 130-134.

Malahotra, S.C. (1996). *Pharmocologial Investigation of Certain Medicinal Plants & Compound Formulation used in Ayurveda and Siddha.* (p. 1). New Delhi: Centre Council for Research in Ayurveda and Siddha.

Maldonado-Celis, M.E., Yahia, E.M., Bedoya, R., Landázuri, P., Loango, N., Aguillón, J., Restrepo, B., Guerrero Ospina, J.C. (2019). Chemical composition of mango (*Mangifera indica* L.) fruit: Nutritional and phytochemical compounds. *Front. Plant Sci., 10*, 1073.
[http://dx.doi.org/10.3389/fpls.2019.01073] [PMID: 31681339]

Malek, F., Boskabady, M.H., Borushaki, M.T., Tohidi, M. (2004). Bronchodilatory effect of *Portulaca oleracea* in airways of asthmatic patients. *J. Ethnopharmacol., 93*(1), 57-62.
[http://dx.doi.org/10.1016/j.jep.2004.03.015] [PMID: 15182905]

Malik, K., Nawaz, F., Nisar, N. (2016). Antibacterial activity of *Amaranthus viridis. Bull Environ Pharmacol Life Sci., 5*, 76-80.

Malik, M., Upadhyay, G. (2021). Leea macrophylla: A Review on ethanobotanical uses, phytochemistry and pharmacological action. *Pharmacogn. Rev., 14*(27), 33-36.
[http://dx.doi.org/10.5530/phrev.2020.14.6]

Manandhar, N.P. (2002). *Plants and People of Nepal..* Portland, OR, USA: Timber Press.

Manchali, S., Chidambara Murthy, K.N., Vishnuvardana, , Patil, B.S. (2021). Vishnuvardana, Patil BS. Nutritional composition and health benefits of various botanical types of melon (*Cucumis melo* L.). *Plants, 10*(9), 1755.
[http://dx.doi.org/10.3390/plants10091755] [PMID: 34579288]

Manda, H., Rao, B.K. (2009). Yashwant, Kutty GN, Swarnkar A, Swarnkar SK. Antioxidant, anti-inflammatory and antipyretic activities of ethyl acetate fraction of ethanolic extract of *Schrebera swietenioides* roxb. root. *Int J Toxicol Pharmacol Res, 1*(1), 7-11.

Manikpuri, N., Jain, S.K., Manoj, K. (2010). Phytochemical investigation of bioactive constituent of some medicinal plants. *Int Res J, 2*(13), 37-38.

Maniyar, Y., Bhixavatimath, P. (2012). Antihyperglycemic and hypolipidemic activities of aqueous extract of *Carica papaya* Linn. leaves in alloxan-induced diabetic rats. *J. Ayurveda Integr. Med., 3*(2), 70-74.
[http://dx.doi.org/10.4103/0975-9476.96519] [PMID: 22707862]

Manju Shree, S., Azamthulla, M. (2019). A review of *Cardiospermum halicacabum* (sapindaceae). *World J. Pharm. Pharm. Sci., 8*(5), 410-420.

Manohari, R.G., Saravanamoorthy, M.D., Vijayakumar, T.P. (2016). Preliminary phytochemical analysis of bamboo seed. *World J. Pharm. Pharm. Sci., 5*, 1336-1342.

Mariani, R., Perdana, F., Fadhlillah, F.M., Qowiyyah, A., Triyana, H. (2019). Antioxidant activity of Indonesian water spinach and land spinach (*Ipomoea aquatica*): A comparative study. *J. Phys. Conf. Ser., 1402*(5), 055091.
[http://dx.doi.org/10.1088/1742-6596/1402/5/055091]

Martín-Cabrejas, M.Á. (2019). *Legumes: Nutritional quality, processing and potential health benefits..* R. Soc. Chem.
[http://dx.doi.org/10.1039/9781788015721]

Martínez, C. (2019). *Canavalia gladiata* and *Dolichos lablab* extracts for sustainable pest biocontrol and plant nutrition improvement in El Salvador. *J Med Plants Stud., 7*(3), 86-93.

Mary, Z., Kumar, K.G., Pasupathy, S., Bikshapathi, T. (2001). Pharmacognostical studies on changeri *Oxalis corniculata* linn. (Oxalidaceae). *Anc. Sci. Life, 21*(2), 120-127.
[PMID: 22557041]

Maseko, I., Mabhaudhi, T., Tesfay, S., Araya, H., Fezzehazion, M., Plooy, C. (2017). African leafy vegetables: a review of status, production and utilization in South Africa. *Sustainability (Basel), 10*(1), 16.
[http://dx.doi.org/10.3390/su10010016]

Mathews, J, Yong, KK, Nurulnahar, BE (2007). Preliminary investigation on biodiversity and its ecosystem in oil palm plantation. Proceedings of Agriculture, Biotechnology & Sustainability Conference, Palm oil: Empowering Change (PIPOC 2007), Malaysian Palm oil Board, Ministry of Plantation industries and commodities, Malaysia; p. 1112-1159.

Mathiventhan, U., Ramiah, S. (2015). Vitamin C content of commonly eaten green leafy vegetables in fresh and under different storage conditions. *Trop. Plant Res., 2*(3), 240-245.

Maundu, P., Ngugi, G., Kabuye, C.H. (1999). *Traditional Food Plants of Kenya, Kenya Resource Centre for Indigenous Knowledge (KENRIK)..* Nairobi, Kenya: National Museum of Kenya.

Mazumdar, T., Hussain, M.S. (2021). A comprehensive review of pharmacological and toxicological properties of *Cheilocostus speciosus* (J. Koenig). *C.D. Specht. Trends Phytochem Res., 5*(1), 2021-12.
[http://dx.doi.org/10.30495/TPR.2021.680480]

McArthur, ED, Stevens, R, Blauer, AC (1983). Growth performance comparisons among 18 accessions of fourwing saltbush (Atriplex canescens) at two sites in central Utah. *Rangeland Ecology & Management/J. Range. Manag. Arch.,.* 36(1): 78-81.
[http://dx.doi.org/10.2307/3897988]

Melbi, B., Hareeshbabu, E. (2018). Pharmacological Evaluation of *Wrightia Tinctoria* – A Review. *IRJPMS, 1*(4), 73-74.

Meng, X., Hu, W., Wu, S., Zhu, Z., Lu, R., Yang, G., Qin, C., Yang, L., Nie, G. (2019). Chinese yam peel enhances the immunity of the common carp (*Cyprinus carpio* L.) by improving the gut defence barrier and modulating the intestinal microflora. *Fish Shellfish Immunol., 95*, 528-537.
[http://dx.doi.org/10.1016/j.fsi.2019.10.066] [PMID: 31678187]

Menon, S., Thakker, J. (2020). Effect of *Amaranthus Viridis* Extract on Growth as Well As Induction of Defense in Chickpea. *Int J Eng Res Tech., 9*(2), 601-603.
[http://dx.doi.org/10.17577/IJERTV9IS020317]

Menon, S.V., Rao, T.V. (2012). Nutritional quality of muskmelon fruit as revealed by its biochemical properties during different rates of ripening. *Int. Food Res. J., 19*(4).

Miller, J.S., Gottschling, M. (2007). Generic classification of *Cordiaceae* (Boraginales). *Syst. Bot., 32*(3), 488-500.

Mirunalini, S., Krishnaveni, M. (2010). Therapeutic potential of *Phyllanthus emblica* (amla): the ayurvedic

wonder. *J Basic Clinical Physiol. Pharma., 21*(1), 93-105.
[http://dx.doi.org/10.1515/jbcpp.2010.21.1.93]

Misar, A.V., Upadhye, A.S., Mujumdar, A.M. (2004). CNS depressant activity of ethanol extract of *Luffa acutangula* Var. Amara CB clarke. *Indian J. Pharm. Sci., 66*, 463-465.

Mishal, H.B., Mishal, R.H. (2020). An Overview of *Ensete superbum*. Chemistry & Pharmacological Profile. *Int J Mod Pharm Res., 4*(4), 18-20.

Mishra, R.K., Upadhyay, V.P., Mohanty, R.C. (2008). Vegetation ecology of the Similipal biosphere reserve, Orissa, India. *Appl. Ecol. Environ. Res., 6*(2), 89-99.
[http://dx.doi.org/10.15666/aeer/0602_089099]

Misra, A., Srivastava, A., Srivastava, S., Rawat, A.K.S. (2016). Simultaneous reverse-phase HPLC determination of major antioxidant phenolics in *Commelina benghalensis* L. tubers. *Acta Chromatogr., 28*(4), 541-554.
[http://dx.doi.org/10.1556/1326.2016.28.3.08]

Misra, B.S. (2005). *Purvardha (Part 1). Bhavaprakasha of Bhavamishra, Shakavarya-10.* (9th ed., pp. 665-836). Varanasi: Chaukamba Sanskrit Samstana.

Misra, L., Wagner, H. (2007). Extraction of bioactive principles from *Mucuna pruriens* seeds. *Indian J. Biochem. Biophys., 44*(1), 56-60.
[PMID: 17385342]

Misra, T.N., Singh, R.S., Pandey, H.S., Singh, B.K., Pandey, R.P. (2001). Constituents of *Asteracantha longifolia*. *Fitoterapia, 72*(2), 194-196.
[http://dx.doi.org/10.1016/S0367-326X(00)00269-0] [PMID: 11223236]

Mitchell, J.D., Pell, S.K., Bachelier, J.B., Warschefsky, E.J., Joyce, E.M., Canadell, L.C., da Silva-Luz, C.L., Coiffard, C. (2022). Neotropical Anacardiaceae (cashew family). *Braz. J. Bot., 45*(1), 139-180.
[http://dx.doi.org/10.1007/s40415-022-00793-5]

Mithen, R., Kibblewhite, H. (1993). *Taxonomy and ecology of Vigna unguiculata (Leguminosae—Papilionoideae) in south-central africa.* Kirkia.

Mohamadpour, M., Khorsandi, L., Mirhoseini, M. (2012). Toxic effect of *Carthamus Tinctorius* L. (Safflower) Extract in the Mouse Testis. 13th Royan Congress on Reproductive Biomedicine and 7th Royan Nursing and Midwifery Seminar; Tehran: Int J Fertil Steril.

Mohammed, H.K., Abraham, A., Saraswathi, R., Mohanta, G.P., Nayar, C. (2012). Formulation and evaluation of herbal gel of *Basella alba* for wound healing activity. *J Pharm Sci Res., 4*, 1642-1648.

Mohan, V.R., Kalidass, C., Amish, D. (2010). Ethno-Medico-Botany of the Palliyars of Saduragiri Hills, Western Ghats, Tamil Nadu. *Econ. Taxon. Bot., 34*, 639-657.

Mohan, V.R., Kalidass, C. (2010). Nutritional and antinutritional evaluation of some unconventional wild edible plants. *Trop. Subtrop. Agroecosystems, 12*(3), 495-506.

Mohite, A.V., Gurav, R.V. (2019). Nutraceutical and antioxidant evaluation of *Abelmoschus* taxa. *Int. J. Veg. Sci., 25*(6), 610-618.
[http://dx.doi.org/10.1080/19315260.2019.1597801]

Mohiuddin AK. Medicinal and therapeutic values of *Sesbania grandiflora*. J Pharm Sci Exp Pharmacol. 2019: 81-6.

[http://dx.doi.org/10.26440/IHRJ/0305.08265]

Mollik, M.A.H., Hossan, M.S., Paul, A.K., Taufiq-Ur-Rahman, M., Jahan, R., Rahmatullah, M. (2010). A Comparative Analysis of Medicinal Plants Used by Folk Medicinal Healers in Three Districts of Bangladesh and Inquiry as to Mode of Selection of Medicinal Plants. *Ethnobot. Res. Appl., 8*, 195-218. [http://dx.doi.org/10.17348/era.8.0.195-218]

Mondal, A. (2014). Phenolic constituents and traditional uses of *Cassia* (Fabaceae) plants: An update. *Org. Biomol. Chem., 3*, 93-141.

Morton, J. (1987) *Fruits of Warm Climates*. J.F. Morton, Miami.

Mostafa, H.A., Elbakry, A.A., Eman, A.A. (2011). Evaluation of antibacterial and antioxidant activities of different plant parts of *Rumex vesicarius* L. (Polygonaceae). *Int. J. Pharm. Pharm. Sci., 3*(2), 109-8.

Motaleb, M.A., Hossain, M.K., Alam, M.K., Mamun, M.M.A.A., Sultana, M. (2013). *Commonly used medicinal herbs and shrubs by traditional herbal practitioners: Glimpses from Thanchi upazila of Bandarban. IUCN.* (pp. 77-78). Dhaka: International Union for Conservation of Nature.

Motaleb, MA, Hossain, MK, Sobhan, I, Alam, MK, Khan, NA, Firoz, R (2011). Selected medicinal plants of Chittagong hill tracts. *International union of conservation of nature and natural resources.*

Mrinal, S., Singh, N. (2018). A review on pharmacological aspects of *Holarrhena antidysenterica. Scholars Academic Journal of Pharmacy., 7*(12), 488-492. [http://dx.doi.org/10.21276/sajp.2018.7.12.5]

Muchate, N.S., Nikalje, G.C., Rajurkar, N.S., Suprasanna, P., Nikam, T.D. (2016). Physiological responses of the halophyte *Sesuvium portulacastrum* to salt stress and their relevance for saline soil bio-reclamation. *Flora (Jena), 224*, 96-105. [http://dx.doi.org/10.1016/j.flora.2016.07.009]

Mudgal, V. (1974). Comparative studies on the anti-inflammatory and diuretic action with different parts of the plant *Boerhaavia diffusa* Linn. (Punarnava). *J. Res. Educ. Indian Med., 9*, 2.

Mukherjee, P.K., Balasubramanian, R., Saha, K., Saha, B.P., Pal, M. (1996). A review on *Nelumbo nucifera* gaertn. *Anc. Sci. Life, 15*(4), 268-276. [PMID: 22556755]

Mulla, S.K., Paramjyothi, S. (2010). Preliminary pharmacognostical and phytochemical evaluation of *Portulaca quadrifida* Linn. *Int. J. Pharm. Tech. Res., 2*(3), 1699-1702.

Murali, A., Ashok, P., Madhavan, V. (2011). *In vitro* antioxidant activity and HPTLC studies on the roots and rhizomes of *Smilax zeylanica* L. (Smilacaceae). *Int. J. Pharm. Pharm. Sci., 3*(1), 192-195.

Muruganandam, A.V., Bhattacharya, S.K., Ghosal, S. (2000). Indole and flavanoid constituents of *Wrightia tinctoria, W. tomentosa* and *W. coccinea. Indian J. Chem., 39*, 125-131.

Muthukumar, T., Udaiyan, K. (2006). Growth of nursery-grown bamboo inoculated with arbuscular mycorrhizal fungi and plant growth promoting rhizobacteria in two tropical soil types with and without fertilizer application. *New For., 31*(3), 469-485. [http://dx.doi.org/10.1007/s11056-005-1380-z]

Muzila, M. (2006). Boerhaavia diffusa L.*Prota 11(1): Medicinal plants/ Plantes médicinales 1..* Wageningen, Netherlands: PROTA. [CD-Rom]

Mwangi, R.W., Macharia, J.M., Wagara, I.N., Bence, R.L. (2021). The medicinal properties of *Cassia fistula* L: A review. *Biomed. Pharmacother., 144*, 112240.
[http://dx.doi.org/10.1016/j.biopha.2021.112240] [PMID: 34601194]

Myerscough, P.J. (1980). *Epilobium Angustifolium* L. *J. Ecol., 68*(3), 1047-1074.
[http://dx.doi.org/10.2307/2259474]

Nabavi, S.F., Habtemariam, S., Jafari, M., Sureda, A., Nabavi, S.M. (2012). Protective role of gallic acid on sodium fluoride induced oxidative stress in rat brain. *Bull. Environ. Contam. Toxicol., 89*(1), 73-77.
[http://dx.doi.org/10.1007/s00128-012-0645-4] [PMID: 22531840]

Nadkarni, K.M., Nadkarni, K.M. (1976). *Indian Materia Medica.* (Vol. I & II). Bombay, India: Popular Prakashan Private Limited.

Nadkarni, K.M. (2009). Indian materia medica. *Bombay Popular Prakashan, 1*, 385-386.

Nagare, S., Deokar, G.S., Nagare, R., Phad, N. (2015). Review on *Coccinia grandis* (L.) voigt (Ivy Gourd). *World J. Pharm. Res., 4910*, 728-743.

Nagarkatti, D.S., Rege, N.N., Desai, N.K., Dahanukar, S.A. (1994). Modulation of Kupffer cell activity by *Tinospora cordifolia* in liver damage. *J. Postgrad. Med., 40*(2), 65-67.
[PMID: 8737554]

Nair, S.L., Chopra, R.N. (1996). Chopra RN. Nat Inst Sci Commun. New Delhi (Pub). Glossary of Indian medicinal Plant 1996; p. 107.

Nair, S.V.G., Hettihewa, M., Rupasinghe, H.P.V. (2014). Apoptotic and inhibitory effects on cell proliferation of hepatocellular carcinoma HepG2 cells by methanol leaf extract of *Costus speciosus. BioMed Res. Int., 2014*, 1-10.
[http://dx.doi.org/10.1155/2014/637098] [PMID: 24818148]

Nakade, D., Mahesh, S., Kiran, N., Vinayak, S. (2013). Phytochemical screening and antibacterial activity of western region wild leaf *Colocasia esculenta. Int. Res. J. Biol. Sci., 2*(10), 18-21.

Nandikar, M.D., Gurav, R.V. (2014). A revision of the genus *Cyanotis* D. Don (Commelinaceae) in India. *Taiwania, 59*(4), 292-314.

Narayanan, N., Thirugnanasambantham, P., Viswanathan, S., Vijayasekaran, V., Sukumar, E. (1999). Antinociceptive, anti-inflammatory and antipyretic effects of ethanol extract of *Clerodendron serratum* roots in experimental animals. *J. Ethnopharmacol., 65*(3), 237-241.
[http://dx.doi.org/10.1016/S0378-8741(98)00176-7] [PMID: 10404422]

Natarajan, D., Britto, S.J., Srinivasan, K., Nagamurugan, N., Mohanasundari, C., Perumal, G. (2005). Anti-bacterial activity of Euphorbia fusiformis—A rare medicinal herb. *J. Ethnopharmacol., 102*(1), 123-126.
[http://dx.doi.org/10.1016/j.jep.2005.04.023] [PMID: 16159702]

Natural Resources Conservation Service – USDA. (2021).

Naudin. In: Ann. Sci. Nat. 1866; Ser. V. 6: 10.

Nayak, P., Thirunavoukkarasu, M. (2016). A review of the plant *Boerhaavia diffusa*: its chemistry, pharmacology and therapeutical potential. *J Phytopharmacol., 5*(2), 83-92.
[http://dx.doi.org/10.31254/phyto.2016.5208]

Nazreen, S. (2022). Ethnomedicinal uses, Pharmacological and Phytochemical Studies of *Bambusa arundinaceae* Retz (A Review). *Orient. J. Chem., 38*(2), 247-258.
[http://dx.doi.org/10.13005/ojc/380204]

Neelam, C., Ranjan, B., Komal, S., Nootan, C. (2011). Review on *Cassia fistula. Int. J. Res. Ayurveda Pharm., 2*, 426-430.

Neharkar, V.S., Pandhare, R. (2015). Acute toxicity study of *Hygrophila auriculata* L. leaves methanolic extract in albino rats. *J Pharm Chem Biol Sci., 3*(3), 388-395.

Ngo Ngwe, MF, Omokolo, DN, Joly, S (2015). Evolution and phylogenetic diversity of yam species (Dioscorea spp.): Implication for conservation and agricultural practices. PLoS one. 21; 10(12): e0145364.
[http://dx.doi.org/10.1371/journal.pone.0145364]

Ngongolo, K., Mtoka, S., Mahulu, A. (2014). The Abundance and pollinators' impact on seed setting of *Leucaena leucocephala* in Wazo Hill restored Quarry, Tanzania. *J Zool Bio Res., 1*(2), 6-10.

Niazi, J., Singh, P., Bansal, Y., Goel, R.K. (2009). Anti-inflammatory, analgesic and antipyretic activity of aqueous extract of fresh leaves of *Coccinia indica. Inflammopharmacology, 17*(4), 239-244.
[http://dx.doi.org/10.1007/s10787-009-0010-3]

Nicolson, D.H. (1979). Nomenclature of *Bombax Ceiba* (Bombacaceae) and *Cochlospermum* (Cochlospermaceae) and their type species. *Taxon, 28*(4), 367-373.
[http://dx.doi.org/10.2307/1219749]

Nikalje, G.C., Shrivastava, M., Nikam, T.D., Suprasanna, P. (2022). Physiological Responses and Tolerance of Halophyte *Sesuvium portulacastrum* L. to Cesium. *Adv. Agric., 2022*, 1-7.
[http://dx.doi.org/10.1155/2022/9863002]

Nikalje, G.C., Srivastava, A.K., Pandey, G.K., Suprasanna, P. (2018). Halophytes in biosaline agriculture: Mechanism, utilization, and value addition. *Land Degrad. Dev., 29*(4), 1081-1095. b
[http://dx.doi.org/10.1002/ldr.2819]

Nikalje, G.C., Variyar, P.S., Joshi, M.V., Nikam, T.D., Suprasanna, P. (2018). Temporal and spatial changes in ion homeostasis and accumulation of flavanoids and glycolipid in a halophyte *Sesuvium portulacastrum* (L.) L. *PLoS One, 13*(4) a
[http://dx.doi.org/10.1371/journal.pone.0193394] [PMID: 29641593]

Nimbal, S.K., Venkatrao, N., Ladde, S., Pujar, B. (2011). Anxiolytic evaluation of *Benincasa hispida* (Thunb) Cogn. fruit extracts. *Int J Pharm Pharm Sci Res., 1*(3), 93-97.

Nirmal, S.A., Dhasade, V.V., Laware, R.B., Rathi, R.A., Kuchekar, B.S. (2011). Antihistaminic effect of *Bauhinia racemosa* leaves. *J. Young Pharm., 3*(2), 129-131.
[http://dx.doi.org/10.4103/0975-1483.80301] [PMID: 21731358]

Nirmal, S.A., Ingale, J.M., Pattan, S.R., Bhawar, S.B. (2013). *Amaranthus roxburghianus* root extract in combination with piperine as a potential treatment of ulcerative colitis in mice. *J. Integr. Med., 11*(3), 206-212.
[http://dx.doi.org/10.3736/jintegrmed2013022] [PMID: 23570686]

Nirmala, A., Gayathri, D.G. (2011). Antidiabetic Activity of *Basella rubra* and its Relationship with the Antioxidant Property. *Br. Biotechnol. J., 1*(1), 1-9.
[http://dx.doi.org/10.9734/BBJ/2011/223]

Nisha, P., Singhal, R.S., Pandit, A.B. (2004). A study on degradation kinetics of ascorbic acid in *amla* (*Phyllanthus emblica L.*) during cooking. *Int. J. Food Sci. Nutr., 55*(5), 415-422.
[http://dx.doi.org/10.1080/09637480412331321823] [PMID: 15545050]

Nissanka, M.C., Weerasekera, M.M., Dilhari, A., Dissanayaka, R., Rathnayake, S., Wijesinghe, G.K. (2023). Phytomedicinal properties of *Hygrophila schulli* (Neeramulliya). *Iran. J. Basic Med. Sci., 26*(9), 979-986.
[PMID: 37605731]

Noda, Y., Kaneyuki, T., Mori, A., Packer, L. (2002). Antioxidant activities of pomegranate fruit extract and its anthocyanidins: delphinidin, cyanidin, and pelargonidin. *J. Agric. Food Chem., 50*(1), 166-171.
[http://dx.doi.org/10.1021/jf0108765] [PMID: 11754562]

Nugroho, A., Heryani, H., Choi, J.S., Park, H.J. (2017). Identification and quantification of flavonoids in *Carica papaya* leaf and peroxynitrite-scavenging activity. *Asian Pac. J. Trop. Biomed., 7*(3), 208-213.
[http://dx.doi.org/10.1016/j.apjtb.2016.12.009]

Ogunmoyole, T., Awodooju, M., Idowu, S., Daramola, O. (2020). *Phyllanthus amarus* extract restored deranged biochemical parameters in rat model of hepatotoxicity and nephrotoxicity. *Heliyon, 6*(12), e05670.
[http://dx.doi.org/10.1016/j.heliyon.2020.e05670] [PMID: 33364479]

O'Keefe, S.J.D. (2016). Diet, microorganisms and their metabolites, and colon cancer. *Nat. Rev. Gastroenterol. Hepatol., 13*(12), 691-706.
[http://dx.doi.org/10.1038/nrgastro.2016.165] [PMID: 27848961]

Okello, J., Ssegawa, P. (2007). Medicinal plants used by communities of Ngai Subcounty, Apac District, northern Uganda. *Afr. J. Ecol., 45*(s1), 76-83.
[http://dx.doi.org/10.1111/j.1365-2028.2007.00742.x]

Okonwu, K., Ekeke, C., Mensah, S. (2017). Ekeke and Mensah SI: Micromorphological and phytochemical studies on *Cleome rutidosperma* Linn. *J. Adv. Biol. Biotechnol., 11*(3), 1-8.
[http://dx.doi.org/10.9734/JABB/2017/31028]

Okuda, T. (2005). Systematics and health effects of chemically distinct tannins in medicinal plants. *Phytochemistry, 66*(17), 2012-2031.
[http://dx.doi.org/10.1016/j.phytochem.2005.04.023] [PMID: 15982679]

Ola, S.S., Catia, G., Marzia, I., Francesco, V.F., Afolabi, A.A., Nadia, M. (2009). HPLC/DAD/MS characterisation and analysis of flavonoids and cynnamoil derivatives in four Nigerian green-leafy vegetables. *Food Chem., 115*(4), 1568-1574.
[http://dx.doi.org/10.1016/j.foodchem.2009.02.013]

Omar, A.S., Nasraldeen, R.A., Albiheyri, R., Kadi, R.H., Abo-Aba, S.E. (2022). The Effects of *Costus speciosus* Root Extract on Cultured Human Lung Cancer Cells, A549. *Biosci. Biotechnol. Res. Commun., 15*(1), 164-170.
[http://dx.doi.org/10.21786/bbrc/15.1.25]

Orni, P.R., Shetu, H.J., Khan, T., Rashed, S.S., Dash, P.R. (2018). A comprehensive review on *Commelina benghalensis* L. (Commelinaceae). *Int J Phacog., 5*(10), 637-645.
[http://dx.doi.org/10.13040/IJPSR.0975-8232.IJP.5(10).637-45]

Ouédraogo, S. (2003). Impact économique des variétés améliorées du niébé sur les revenus des exploitations agricoles du plateau central du Burkina Faso. *Tropicultura, 21*, 204-210.

Oviya, I.R., Sharanya, M., Jeyam, M. (2015). Phytochemical and pharmacological assessment of *Wrightia tinctoria* R. BR. A review. World. *J. Pharma Res., 4*, 1992-2015.

Paiva, S.R., Figueiredo, M.R., Aragão, T.V., Kaplan, M.A.C. (2003). Antimicrobial activity *in vitro* of plumbagin isolated from *Plumbago* species. *Mem. Inst. Oswaldo Cruz, 98*(7), 959-961. [http://dx.doi.org/10.1590/S0074-02762003000700017] [PMID: 14762525]

Padhan, A.R., Agrahari, A.K., Meher, A. (2010). A study on antipyretic activity of *Capparis zeylanica* Linn. Plant methanolic extract. *Int. J. Pharm. Sci. Res., 1*(3), 169-171.

Prasad, M.S., Ramachandran, A., Chandola, H., Harisha, C.R., Shukla, V. (2012). Pharmacognostical and phytochemical studies of *Curcuma neilgherrensis* (Wight) leaf - A folklore medicine. *Ayu, 33*(2), 284-288. [http://dx.doi.org/10.4103/0974-8520.105253] [PMID: 23559805]

Panchawat, S. (2012). *Ficus religiosa* Linn. (Peepal): A Phyto-Pharmacological Review. *Inter J Pharm Chem Sci., 1*(1), 435-446.

Panda, B.K., Patro, V.J., Mishra, U.S. Subratkar. (2014) Comparative Study of Anti-Pyretic Activity between Acetone and Ethanol Stem Bark Extracts of *Spondias pinnata* (Linn.F) Kurz. *Res. J. Pharm. Biol. Chem. Sci., 1*(1), 26-32.

Panda, H. (2002). *Medicinal plants cultivation and their uses..* Delhi: Asia Pacific Business Press Inc..

Panda, P., Das, D., Dash, P., Ghosh, G. (2015). Therapeutic potential of *Bauhinia racemosa*—a mini review. *Int. J. Pharm. Sci. Rev. Res., 32*(2), 169-179.

Pande, M., Pathak, A. (2010). Preliminary pharmacognostic evaluations and phytochemical studies on leaf of *Chenopodium album* (Bathua Sag). *Asian J. Exp. Biol. Sci., 1*(1), 91-95.

Pandey, A., Ranjan, P., Ahlawat, S.P., Bhardwaj, R., Dhariwal, O.P., Singh, P.K., Malav, P.K., Harish, G.D., Prabhu, P., Agrawal, A. (2021). Studies on fruit morphology, nutritional and floral diversity in less-known melons (*Cucumis melo* L.) of India. *Genet. Resour. Crop Evol., 68*(4), 1453-1470. [http://dx.doi.org/10.1007/s10722-020-01075-3]

Pandey, A., Verma, R.K. (2018). Taxonomical and pharmacological status of *Typha*: A Review. *Ann. Plant Sci., 7*(3), 2101-2106. [http://dx.doi.org/10.21746/aps.2018.7.3.2]

Pandey G. (2003). Dravyaguna Vijnana. Varanasi: Krishnadas Academy 2003; vol (3) 728-32.

Pandey, V.C., Kumar, A. (2013). *Leucaena leucocephala*: an underutilized plant for pulp and paper production. *Genet. Resour. Crop Evol., 60*(3), 1165-1171. [http://dx.doi.org/10.1007/s10722-012-9945-0]

Pardeshi, S.K., Vikhe, D.N., Ghawate, V.B., Bhone, V.V. (2023). Phytochemical investigation and pharmacological assessment of *Caralluma adscendens* (ROXB.). *Eur. Chem. Bull., 12*(10), 717-722.

Parekh, J., Chanda, S. (2007). *In vitro* screening of antibacterial activity of aqueous and alcoholic extracts of various Indian plant species against selected pathogens from Enterobacteriaceae. *Afr. J. Microbiol. Res., 1*(6), 92-99.

Parrota, J.A. (2001). Haeling plants of Pennninsular India, CABI publishing, CAB international Newyork, USA, 56.

Parveen, U.B., Upadhyay, B., Roy, S., Kumar, A. (2007). Traditional uses of medicinal plants among the rural communities of Churu district in the Thar Desert, India. *J. Ethnopharmacol., 113*(3), 387-399.

[http://dx.doi.org/10.1016/j.jep.2007.06.010] [PMID: 17714898]

Patel, A., Singh, J. (2023). Taro (*Colocasia esculenta* L): Review on Its botany, morphology, ethno medical uses, Phytochemistry and pharmacological activities. *Pharma Innov., 12*(3), 05-14.
[http://dx.doi.org/10.22271/tpi.2023.v12.i3a.18908]

Patel, A.K., Pathak, N., Trivedi, H., Gavania, M., Patel, M., Panchal, N. (2011). Phytopharmacological properties of *Cordia dichotoma* as a potential medicinal tree: an overview. *Int J Inst Pharm Life Sci., 1*(1), 40-51.

Patel, J.R., Tripathi, P., Sharma, V., Chauhan, N.S., Dixit, V.K. (2011). *Phyllanthus amarus*: Ethnomedicinal uses, phytochemistry and pharmacology: A review. *J. Ethnopharmacol., 138*(2), 286-313.
[http://dx.doi.org/10.1016/j.jep.2011.09.040] [PMID: 21982793]

Patel, DK (2018). *Diplocyclos palmatus* (L.) Jeffry: Morphological variations and medicinal values. *J. Med. Plants Stud., 6*(1), 3-5.

Patil, MV, Patil, DA (2005). Ethnomedicinal practices of Nasik district, Maharashtra. 4(3): 287-290

Patiño-Rodríguez, O., Agama-Acevedo, E., Pacheco-Vargas, G., Alvarez-Ramirez, J., Bello-Pérez, L.A. (2019). Physicochemical, microstructural and digestibility analysis of gluten-free spaghetti of whole unripe plantain flour. *Food Chem., 298*, 125085.
[http://dx.doi.org/10.1016/j.foodchem.2019.125085] [PMID: 31260951]

Paudel, K.R., Panth, N. (2015). Phytochemical profile and biological activity of *Nelumbo nucifera*. *Evid. Based Complement. Alternat. Med., 2015*, 1-16.
[http://dx.doi.org/10.1155/2015/789124] [PMID: 27057194]

Pawar, V.A., Pawar, P.R. (2014). *Costus speciosus*: an important medicinal plant. *Int. J. Sci. Res., 3*(7), 28-32.

Pekamwar, S.S., Kalyankar, T.M., Kokate, S.S. (2013). Pharmacological Activities of *Coccinia grandis*. *J. Appl. Pharm. Sci., 3*(5), 114-119. [Review].
[http://dx.doi.org/10.7324/JAPS.2013.3522]

Perry, L.M. (1980). *Medicinal Plants of East and Southeast Asia.* (p. 211). Cambridge: MIT Press.

Philcox, D. (1997). A Revised Handbook to the FLORA OF CEYLON, M. D. Dassanayake, W. D. Clayton. New Delhi: Oxford & IBH Publishing; Cucurbitaceae p. 2630.

Pillai, L.S., Nair, B.R. (2012). Pharmacognostical standardization and phytochemical studies in *Cleome viscosa* L. and *Cleome burmanii* W. & A. (Cleomaceae). *J. Pharm. Res., 5*(2), 1231-1235.

Pitrat, M. (2017). Melon genetic resources: phenotypic diversity and horticultural taxonomy. Genetics and genomics of Cucurbitaceae: 25-60.

Plunkett, GM, Pimenov, MG, Reduron, JP, Kljuykov, EV, van Wyk, BE, Ostroumova, TA, Henwood, MJ, Tilney, PM, Spalik, K, Watson, MF, Lee, BY (2018). Apiaceae: Umbelliferae Juss., Gen. Pl.: 218 (1789), nom. cons. et nom. alt. Flowering plants. Eudicots: apiales, gentianales (except rubiaceae). 9-206.

Pojar J, MacKinnon A, Card G, Foraging PN, from Alaska E, Guide TP, Guide F. (1994). Plants of the Pacific Northwest. British Columbia Ministry of Forest and Lone Pine Publishing, BC, Canada.

Pooja, L.V., Verma, A. (2017). GC-MS and phytopharmacological analysis of aqueous distillate of *Boerhavia diffusa* roots. *Int. J. Pharm. Pharm. Res., 10*, 374-391.

Poonia, A., Upadhayay, A. (2015). *Chenopodium album* Linn: review of nutritive value and biological properties. *J. Food Sci. Technol., 52*(7), 3977-3985.
[http://dx.doi.org/10.1007/s13197-014-1553-x] [PMID: 26139865]

Poonia, A., Upadhayay, A. (2015). *Chenopodium album* Linn: review of nutritive value and biological properties. *J. Food Sci. Technol., 52*(7), 3977-3985.
[http://dx.doi.org/10.1007/s13197-014-1553-x] [PMID: 26139865]

Prabhakar, Y.S., Suresh, K.D. (1990). A survey of cardioactive drug formulations from Ayurveda. II: porridges, oils, clarified butters, electuaries, pastes, ash preparations and calcined powders. *Fitoterapia, 61*, 395-416.

Pradeep Kumar, C.H. (2014). Narsimha Reddy. Protective effect of *Canavalia gladiata* (sword bean) fruit extracts and its flavanoidal contents, against azathioprine-induced toxicity in hepatocytes of albino rats. *Toxicol. Environ. Chem., 96*(3), 474-481.
[http://dx.doi.org/10.1080/02772248.2014.950805]

Prajapati ND, Purohit SS, Sharma AK, Kumar T. (2003). A handbook of medicinal plants: A complete source book. In A handbook of medicinal plants: a complete source book; pp. 554-554.

Prakash, G., Hosetti, B.B. (2012). Bio-efficacy of *Dioscorea pentaphylla* from Midmid-Western Ghats, India. *Toxicol. Int., 19*(2), 100-105.
[http://dx.doi.org/10.4103/0971-6580.97195] [PMID: 22778504]

Prananda, AT, Dalimunthe, A, Harahap, U (2023). Phyllanthus emblica: a comprehensive review of its phytochemical composition and pharmacological properties. *Front. Pharmacol.* 26; 14: 1288618.
[http://dx.doi.org/10.3389/fphar.2023.1288618]

Prasad, K.J.R., Gopinath, K.R., Reddy, H.V., Harini, S.S., Ramakrishnan, R. (2019). Isolation of new Volatile ester and Ellagitannin from *Glossocardia bosvallia* (L.f) DC-Asteraceae. *Appl. Chem., 8*(6), 2386-2392.

Prasad, K.N., Shivamurthy, G.R., Aradhya, S.M. (2008). *Ipomoea aquatica*, an underutilized green leafy vegetable: a review. *Int. J. Bot., 4*(1), 123-129.
[http://dx.doi.org/10.3923/ijb.2008.123.129]

Prince, P.S., Kamalakkannan, N., Menon, V.P. (2004). Restoration of antioxidants by ethanolic *Tinospora cordifolia* in alloxan-induced diabetic Wistar rats. *Acta Pol. Pharm., 61*(4), 283-287.
[PMID: 15575595]

Patel, J.J., Acharya, S.R., Acharya, N.S. (2014). *Clerodendrum serratum* (L.) Moon. – A review on traditional uses, phytochemistry and pharmacological activities. *J. Ethnopharmacol., 154*(2), 268-285.
[http://dx.doi.org/10.1016/j.jep.2014.03.071] [PMID: 24727551]

Pullaiah, T. (2006). *Encyclopedia of world medicinal plants. Daya books.*

Pal, M., Rawat, P., Saroj, L.M., Kumar, A., Singh, T.D., Tewari, S.K. (2016). Phytochemicals and cytotoxicity of *Launaea procumbens* on human cancer cell lines. *Pharmacogn. Mag., 12*(47) (Suppl. 4), 431.
[http://dx.doi.org/10.4103/0973-1296.191452] [PMID: 27761070]

Pungulani, L.L., Millner, J.P., Williams, W.M., Banda, M. (2013). Improvement of leaf wilting scoring system in cowpea (*Vigna unguiculata* (L) Walp. from qualitative scale to quantitative index. *Aust. J. Crop Sci., 7*, 1262.

Qaiser, M., Jafri, S.M.H. (1975). Commelina benghalensis. *Flora of Pak., 84*, 10.

Qamar, S., Shaikh, A. (2018). Therapeutic potentials and compositional changes of valuable compounds from banana- A review. *Trends Food Sci. Technol., 79*, 1-9.
[http://dx.doi.org/10.1016/j.tifs.2018.06.016]

Quasim, C., Dutta, N.L. (1967). Presence of stigmasterol in the root of *Asteracantha longifolia* Nees. *J. Indian Chem. Soc., 44*, 82-83.

Quazi, M.A., Molvi, K.I. (2014). Ethnomedicinal survey of *Meyna laxiflora* in tribes of Akkalkuwa, Nandurbar District. *Int. J. Pharma Sci., 5*(3), 225-230.

Quézel P, Santa S, Schotter O, Emberger L. (1962). Nouvelle flore de l'Algérie et des régions désertiques méridionales. Éditions du Centre national de la Recherche scientifique, Paris; 7e.

Qureshi, S.J., Khan, M.A. (2000). Ethnobotanical study of Kahuta from Rawalpindi district Pakistan. *J. Biol. Sci. (Faisalabad, Pak.), 1*(1), 27-30.
[http://dx.doi.org/10.3923/jbs.2001.27.30]

Rosarlo, M., Vinas, A., Lorenz, K. (1999). Pasta products containing taro (*Colocasia esculenta* l. Schott) and Chaya (*Cnidoscolus chavamansa* L. mcvaugh). *J. Food Process. Preserv., 23*(1), 1-20.
[http://dx.doi.org/10.1111/j.1745-4549.1999.tb00366.x]

Raghunathan, M. (2017). An ethnomedicinal survey of medicinal plants utilized by folk people of the Thrissur Forest circle, Kerala. *Eur. J. Pharm. Med. Res., 4*, 401-409.

Rahimi, V.B., Ajam, F., Rakhshandeh, H., Askari, V.R. (2019). A pharmacological review on *Portulaca oleracea* L.: focusing on anti-inflammatory, anti-oxidant, immuno-modulatory and antitumor activities. *J. Pharmacopuncture, 22*(1), 7-15.
[http://dx.doi.org/10.3831/KPI.2019.22.001] [PMID: 30988996]

Ruoxi YIN, Wei HE, Barbara GÓRNA, Roman HOŁUBOWICZ. (2017) Seed Yield and Quality of Sword Bean (*Canavalia gladiata* (Jacq.) DC.) produced in Poland. *Not Bot Horti Agrobo, 45*(2), 561-568.
[http://dx.doi.org/10.15835/nbha45210888]

Rahman, M.A., Imran, T., Islam, S. (2013). Antioxidative, antimicrobial and cytotoxic effects of the phenolics of *Leea indica* leaf extract. *Saudi J. Biol. Sci., 20*(3), 213-225.
[http://dx.doi.org/10.1016/j.sjbs.2012.11.007] [PMID: 23961238]

Rajagopal, P.L., Premaletha, K., Kiron, S.S., Sreejith, K.R. (2013). Phytochemical and pharmacological review on *Cassia fistula* Linn. the golden shower. *Int. J. Pharm. Chem. Biol. Sci., 3*, 672-679.

Rajagopal, P.L., Premaletha, K., Sreejith, K.R. (2016). Anthelmintic activity of the flowers of *Sesbania grandiflora*. *Pers. J. Innov. Appl. Pharm. Sci., 1*, 8-11.

Rajaputana, L.M., Divya, M.S.S., Bhavani, B., Sudhakar, M. (2022). A review on *Amaranthus roxburghianus* nevski. *World J. Pharm. Pharm. Sci., 11*(11), 1451-1464.

Rajaram, N., Janardhanan, K. (1992). Nutritional and chemical evaluation of raw seeds of*Canavalia gladiata* (Jacq) DC. and *C.ensiformis* DC: The under utilized food and fodder crops in India. *Plant Foods Hum. Nutr., 42*(4), 329-336.
[http://dx.doi.org/10.1007/BF02194094] [PMID: 1438077]

Rajasab, A.H., Isaq, M. (2004). Documentation of folk knowledge on edible wild plants of north Karnataka.

Int J Tradit. Knowl., 3(4), 419-429.

Rajasekaran, A., Sivakumar, V., Darlinquine, S. (2012). Evaluation of wound healing activity of *Ammannia baccifera* and *Blepharis maderaspatensis* leaf extracts on rats. *Rev. Bras. Farmacogn., 22*(2), 418-427. [http://dx.doi.org/10.1590/S0102-695X2011005000207]

Rajesh, P., Selvamani, P., Latha, S., Saraswathy, A., Kannan, V.R. (2009). A review on chemical and medicobiological applications of capparidaceae family. *Phcog. Rev., 3*, 378-387.

Rajesh, V., Perumal, P. (2014). *In-vitro* cytoprotective activity of *Smilax zeylanica* leaves against hydrogen peroxide induced oxidative stress in L-132 and BRL 3A cells. *Orient. Pharm. Exp. Med., 14*(3), 255-268. [http://dx.doi.org/10.1007/s13596-014-0154-6]

Rajeshwar, Y., Kumar, S.G.P., Gupta, M., Mazumder, K.U. (2005). Studies on *in vitro* antioxidant activities of mhetanol extract of *Mucuna pruriens* (Fabaceae) seeds. *Eur Bull Drug Res., 13*, 31-39.

Rajlakshmi, D., Banerjee, S.K., Sood, S., Maulik, S.K. (2010). *In-vitro* and *in-vivo* antioxidant activity of different extracts of the leaves of *Clerodendron colebrookianum* Walp in the rat. *J. Pharm. Pharmacol., 55*(12), 1681-1686. [http://dx.doi.org/10.1211/0022357022296] [PMID: 14738596]

Rajopadhye, A.A., Upadhye, A.S., Taware, S.P. (2015). Bioactivity of indigenous plant *Glossocardia bosvallia* (Lf) DC. against insect pests of stored products. *Indian J. Tradit. Knowl., 15*(2), 260-265.

Raju, S., Kavimani, S., Uma Maheshwara rao, V., Sreeramulu Reddy, K., Vasanth Kumar, G. (2011). Floral extract of *Tecoma stans*: A potent inhibitor of gentamicin-induced nephrotoxicity *in vivo. Asian Pac. J. Trop. Med., 4*(9), 680-685. b [http://dx.doi.org/10.1016/S1995-7645(11)60173-9] [PMID: 21967688]

Raju, S., Kavimani, S., Uma, M.R., Sreeramulu, R.K. (2011). *Tecoma stans* (L.) Juss. ex Kunth (Bignoniaceae): Ethnobotany, phytochemistry and pharmacology. *J. Pharm. Biomed. Sci., 8*, 1-5. a

Ramakrishnan, R., Samydurai, P., Thangapandian, V. (2013). Qualitative Phytochemical Analysis and *In Vitro* Antibacterial activity of *Glossocardia bosvallia* (Linn. f.) Dc.-An Important Ayurvedic Medicinal Herb. *Res J Pharm Technol., 6*(12), 1391-1396.

Ramandeep, K., Nahid, A., Neelabh, C., Navneet, K. (2017). Phytochemical screening of *Phyllanthus niruri* collected from Kerala region and its antioxidant and antimicrobial potentials. *J Pharm Sci Res., 9*(8), 1312.

Ramesh, M., Gayathri, V., Rao, A.A., Prabhakar, M.C., Rao, C.S. (1989). Pharmacological actions of fruit juice of *Benincasa hispida. Fitoterapia, 60*(3), 241-247.

Rana, S., Suttee, A. (2012). Phytochemical investigation and evaluation of free radical scavenging potential of *Benincasa hispida* peel extracts. *Int. J. Curr. Pharm. Rev. Res., 3*(3), 43-46.

Rao, D.S., Rao, G.M.N. (2015). Sacred grove of Punyagiri Hill, Vizianagaram District, Ap, India: ecological and sociological study. *Int. J. Environ., 4*(1), 30-47. [http://dx.doi.org/10.3126/ije.v4i1.12176]

Rao, N.V., Prakash, K.C., Kumar, S.M. (2006). Pharmacological investigation of *Cardiospermum halicacabum* Linn in different animal models of diarrhoea. *Int. J. Pharmacol., 38*, 346-349. [http://dx.doi.org/10.4103/0253-7613.27703]

Rasal, A.S., Nayak, P.G., Baburao, K., Shenoy, R.R., Mallikarjuna Rao, C. (2009). Evaluation of the healing potential of *Schrebera swietenioides* in the dexamethasone-suppressed wound healing in rodents. *Int. J. Low.*

Extrem. Wounds, 8(3), 147-152.
[http://dx.doi.org/10.1177/1534734609344208] [PMID: 19703950]

Rashed, K., Butnariu, M. (2014). Antimicrobial and antioxidant activities of *Bauhinia racemosa* Lam. and chemical content. *Iran. J. Pharm. Res., 13*(3), 1073-1080.
[PMID: 25276210]

Rashmi, D.R., Raghu, N., Gopenath, T.S. (2018). Taro (*Colocasia esculenta*): An overview review. *J. Med. Plants Stud., 6*(4), 156-161.

Rasul, M.G., Hiramatsu, M., Okubo, H. (2004). Morphological and physiological variation in kakrol (*Momordica dioica* Roxb.). *J. Fac. Agric. Kyushu Univ., 49*(1), 1-11.
[http://dx.doi.org/10.5109/4559]

Rathod, M.C., Dhale, D. (2014). Pharmacognostic characterization and phytochemical screening of *Launaea procumbens*. *Int. J. Pharm. Res. Sch., 3*, 41-50.

Ratz-Łyko, A., Arct, J., Majewski, S., Pytkowska, K. (2015). Influence of polyphenols on the physiological processes in the skin. *Phytother. Res., 29*(4), 509-517.
[http://dx.doi.org/10.1002/ptr.5289] [PMID: 25586195]

Reddy, Y.S.R., Venkatesh, S., Ravichandran, T., Subburaju, T., Suresh, B. (1999). Pharmacognostical studies on *Wrightia tinctoria* bark. *Pharm. Biol., 37*(4), 291-295.
[http://dx.doi.org/10.1076/phbi.37.4.291.5798]

Redondo-Gómez, S., Mateos-Naranjo, E., Figueroa, M.E., Davy, A.J. (2010). Salt stimulation of growth and photosynthesis in an extreme halophyte, *Arthrocnemum macrostachyum*. *Plant Biol., 12*(1), 79-87.
[http://dx.doi.org/10.1111/j.1438-8677.2009.00207.x] [PMID: 20653890]

Rege, N., Bapat, R.D., Koti, R., Desai, N.K., Dahanukar, S. (1993). Immunotherapy with *Tinospora cordifolia*: a new lead in the management of obstructive jaundice. *Indian J. Gastroenterol., 12*(1), 5-8.
[PMID: 8330924]

Rekha, C. (2015). *Momordica cymbalaria* a nutritious underutilized vegetable taxonomy, nutritional, medicinal, propagation, hybridization and cytological aspects. *Int. J. Agric. Sci., 5*(4), 255-261.

Reyad-ul-Ferdous, M., Shahjahan, D.S., Tanvir, S., Mukti, M. (2015). Present biological status of potential medicinal plant of *Amaranthus viridis*: a comprehensive review. *Am J Clin Med., 3*(5), 12.
[http://dx.doi.org/10.11648/j.ajcem.s.2015030501.13]

Rezaei, J., Rouzbehan, Y., Fazaeli, H., Zahedifar, M. (2013). Carcass characteristics, non-carcass components and blood parameters of fattening lambs fed on diets containing amaranth silage substituted for corn silage. *Small Rumin. Res., 114*(2-3), 225-232.
[http://dx.doi.org/10.1016/j.smallrumres.2013.06.012]

Rinchen, T., Singh, N., Maurya, S.B., Soni, V., Phour, M., Kumar, B. (2017). Morphological characterization of indigenous vegetable (*Atriplex hortensis* L.) from trans-Himalayan region of Ladakh (Jammu and Kashmir), India. *Aust. J. Crop Sci., 11*(3), 258-263.
[http://dx.doi.org/10.21475/ajcs.17.11.03.pne261]

Rivera, N.V., Pérez, S.C., Chanona-Pérez, J.J. (2014). Industrial applications and potential pharmaceutical uses of mango (Mangifera indica) kernel.*Seeds as Functional Foods and Nutraceuticals: New Frontiers in Food Science.* Nova Science Publishers, Inc.

Rodda, R., Kota, A., Sindhuri, T., Kumar, S.A., Gnananath, K. (2012). Investigation on anti-inflammatory property of *Basella alba* L. leaf extract. *Int. J. Pharm. Pharm. Sci.,* *4*, 452-454.

Rojas, A., Hernandez, L., Pereda-Miranda, R., Mata, R. (1992). Screening for antimicrobial activity of crude drug extracts and pure natural products from Mexican medicinal plants. *J. Ethnopharmacol.,* *35*(3), 275-283. [http://dx.doi.org/10.1016/0378-8741(92)90025-M] [PMID: 1548900]

Rook, A.J., Dumont, B., Isselstein, J., Osoro, K., WallisDeVries, M.F., Parente, G., Mills, J. (2004). Matching type of livestock to desired biodiversity outcomes in pastures – a review. *Biol. Conserv.,* *119*(2), 137-150. [http://dx.doi.org/10.1016/j.biocon.2003.11.010]

Roshan, A., Verma, N.K., Gupta, A. (2014). A Brief Study on *Carica papaya*- A Review. *Int. J. Curr. Pharm. Res.,* *2*(4), 541-550.

Ross, I.A. (2003). Hibiscus sabdariffa.*Medicinal Plants of the World.*. Totowa, NJ: Humana Press. [http://dx.doi.org/10.1007/978-1-59259-365-1_13]

Ross, I.A. (2005). *Medicinal plants of the world. Chemical constituents, traditional and modern medicinal uses.* Humana Press Incorporated.

Rout, K. Swain. (2005). Conservation strategy for medicinal plants in Orissa, India. *Int. J. Biodiv Sci Manag.,* *1*(4), 205-211. [http://dx.doi.org/10.1080/17451590509618093]

Roy, A., Bharadvaja, N. (2017). A review on pharmaceutically important medical plant: *Plumbago zeylanica*. *J. Ayurvedic Herb. Med.,* *3*(4), 225-228. [http://dx.doi.org/10.31254/jahm.2017.3411]

Rutuja, R.S., Shivsharan, U., Shruti, A.M. (2015). *Ficus religiosa* (Peepal): A phytochemical and pharmacological review. *Intl J Pharm Chem Sci.,* *4*(3), 360-370.

Rymbai, H., Mawlein, J., Rymbai, D. (2022). *Meyna laxiflora* Robyns a potential multipurpose tree: an underutilized fruit and medicinal tree in Meghalaya. *Agri J World,* *2*(5), 1-7.

Sabandar, C.W., Jalil, J., Ahmat, N., Aladdin, N.A. (2017). Medicinal uses, chemistry and pharmacology of *Dillenia* species (Dilleniaceae). *Phytochemistry,* *134*, 6-25. [http://dx.doi.org/10.1016/j.phytochem.2016.11.010] [PMID: 27889244]

Sadique, J., Chandra, T., Thenmozhi, V., Elango, V. (1987). Biochemical modes of action of *Cassia occidentalis* and *Cardiospermum halicacabum* in inflammation. *J. Ethnopharmacol.,* *19*(2), 201-212. [http://dx.doi.org/10.1016/0378-8741(87)90042-0] [PMID: 3613609]

Saha, S. (2017). *Oxalis corniculata* Linn. (Amrul): Magical plant. *Am Int J Res. in For, Appl. Nat. Sci.,* *18*(1), 16-19.

Singh Gill, N., Arora, R., Kumar, S.R. (2011). Evaluation of antioxidant, anti-inflammatory and analgesic potential of the *Luffa acutangula* Roxb. Var. amra. *Res. J. Phytochem.,* *5*(4), 201-208. [http://dx.doi.org/10.3923/rjphyto.2011.201.208]

Sairam, K., Dorababu, M., Goel, R.K., Bhattacharya, S.K. (2002). Antidepressant activity of standardized extract of *Bacopa monniera* in experimental models of depression in rats. *Phytomedicine,* *9*(3), 207-211. [http://dx.doi.org/10.1078/0944-7113-00116] [PMID: 12046860]

Sajwan, V.S., Harjal, N., Paliwal, G.S. (1977). Developmental anatomy of the leaf of L. *Ficus religiosa*. *Ann. Bot. (Lond.),* *41*(2), 293-302.

[http://dx.doi.org/10.1093/oxfordjournals.aob.a085291]

Sakano, Y., Mutsuga, M., Tanaka, R., Suganuma, H., Inakuma, T., Toyoda, M., Goda, Y., Shibuya, M., Ebizuka, Y. (2005). Inhibition of human lanosterol synthase by the constituents of *Colocasia esculenta* (taro). *Biol. Pharm. Bull., 28*(2), 299-304.
[http://dx.doi.org/10.1248/bpb.28.299] [PMID: 15684488]

Sakia, B., Rawat, J.S., Tag, H., Das, A.K. (2011). An Investigation on the Taxonomy and Ecology of the Genus *Dioscorea* in Arunanchal Pradesh, India. *J Front Res. Arts Sci., 01*, 44-53.

Sambavi, N., Naveena Kumari, K., Balaji, A.A. (2022). Evaluation of *Amaranthus roxiburghianus* Linn Plant extract on Breast Cancer MCF-7 Cells by MTT Assay. *YMER, 21*(12), 137-151.

Sameh, S., Al-Sayed, E., Labib, R.M., Singab, A.N. (2018). Genus *Spondias*: A phytochemical and pharmacological review. *Evid. Based Complement. Alternat. Med., 2018*(1), 5382904.
[http://dx.doi.org/10.1155/2018/5382904] [PMID: 29785194]

Samuelsson, G., Farah, M.H., Claeson, P., Hagos, M., Thulin, M., Hedberg, O., Warfa, A.M., Hassan, A.O., Elmi, A.H., Abdurahman, A.D., Elmi, A.S., Abdi, Y.A., Alin, M.H. (1992). Inventory of plants used in traditional medicine in Somalia. II. Plants of the families Combretaceae to Labiatae. *J. Ethnopharmacol., 37*(1), 47-70.
[http://dx.doi.org/10.1016/0378-8741(92)90004-B] [PMID: 1453703]

Samydurai, P., Jagatheshkumar, S., Aravinthan, V., Thangapandian, V. (2012). Survey of wild aromatic ethnomedicinal plants of Velliangiri Hills in the Southern Western Ghats of Tamil Nadu, India. *Int. J. Med. Aromat. Plants, 2*(2), 229-234.

Saraswat, R., Pokharkar, R. (2012). GC-MS Studies of *Mimosa pudica. Int. J. Pharm. Tech. Res., 4*(1), 93-98.

Sarkar, A., Pradhan, S., Mukhopadhyay, I., Bose, S.K., Roy, S., Chatterjee, M. (1999). Inhibition of early DNA-damage and chromosomal aberrations by *Trianthema portulacastrum*l. In carbon tetrachloride-induced mouse liver damage. *Cell Biol. Int., 23*(10), 703-708.
[http://dx.doi.org/10.1006/cbir.1999.0439] [PMID: 10736194]

Sarkar, T., Ghosh, P., Poddar, S., Choudhury, S., Sarkar, A., Chatterjee, S. (2020). *Oxalis corniculata* Linn. (Oxalidaceae): A brief review. *J. Pharmacogn. Phytochem., 9*(4), 651-655.
[http://dx.doi.org/10.22271/phyto.2020.v9.i4i.11777]

Sarker, U., Islam, M.T., Rabbani, M.G., Oba, S. (2017). Genotypic diversity in vegetable amaranth for antioxidant, nutrient and agronomic traits. *Indian J. Genet. Plant Breed., 77*(1), 173-176.
[http://dx.doi.org/10.5958/0975-6906.2017.00025.6]

Sarker, U., Islam, T., Rabbani, G., Oba, S. (2015). Genotype variability in composition of antioxidant vitamins and minerals in vegetable amaranth. *Genetika, 47*(1), 85-96.
[http://dx.doi.org/10.2298/GENSR1501085S]

Saroj Kumar, V., Phurailatpam, A.K., Jaishanker, R., Annamalai, A. (2012). *Ensete superbum*: A multi-utility plant. *Indian Hortic., 57*, 31.

Sarvalingam, A., Rajendran, A. (2016). Rare, endangered and threatened (RET) climbers of Southern Western Ghats, India. *Rev. Chil. Hist. Nat., 89*(1), 9.
[http://dx.doi.org/10.1186/s40693-016-0058-6]

Sathiyanarayanan, L., Arulmozhi, S. (2007). *Mucuna pruriens* a comprehensive review. *Phacog Rev., 1*,

157-162.

Sathya, E., Muthukumar, T., Sekar, T. (2022). Comparative vegetative anatomy of *Wrightia tinctoria* R.Br. and the endemic *Wrightia indica* Ngan (Apocynaceae Juss.) occurring in peninsular India. *Flora (Jena)*, *290*, 152043.
[http://dx.doi.org/10.1016/j.flora.2022.152043]

Schaefer, H., Renner, S.S. (2011). Phylogenetic relationships in the order Cucurbitales and a new classification of the gourd family (Cucurbitaceae). *Taxon*, *60*(1), 122-138.
[http://dx.doi.org/10.1002/tax.601011]

Schieber, A., Berardini, N., Carle, R. (2003). Identification of flavonol and xanthone glycosides from mango (*Mangifera indica* L. Cv. "Tommy Atkins") peels by high-performance liquid chromatography-electrospray ionization mass spectrometry. *J. Agric. Food Chem.*, *51*(17), 5006-5011.
[http://dx.doi.org/10.1021/jf030218f] [PMID: 12903961]

Sebastian, M.K., Bhandari, M.M. (1984). Medico-ethno botany of Mount Abu, Rajasthan, India. *J. Ethnopharmacol.*, *12*(2), 223-230.
[http://dx.doi.org/10.1016/0378-8741(84)90050-3] [PMID: 6521495]

Seguin, P., Mustafa, A.F., Donnelly, D.J., Gélinas, B. (2013). Chemical composition and ruminal nutrient degradability of fresh and ensiled amaranth forage. *J. Sci. Food Agric.*, *93*(15), 3730-3736.
[http://dx.doi.org/10.1002/jsfa.6218] [PMID: 23653266]

Senniappan, P., Karthikeyan, V., Janarthanan, L., Sathiyabalan, G., Sriram, K. (2020). *Merremia emarginata* Burm. F: Pharmacognostical Standardization and Phytochemical Studies of Its Leaves. *GIS Sci J*, *7*(11), 117-127.

Senthilkumar, S., Vijayakumari, K. (2013). Phytochemical and GC-MS analysis of *Cardiospermum halicacabum* Linn. Leaf. *Int J Inst Pharm Life Sci.*, *3* (Suppl. 5), 95-98.

Sesin, V., Davy, C.M., Freeland, J.R. (2021). Review of *Typha* spp. (cattails) as toxicity test species for the risk assessment of environmental contaminants on emergent macrophytes. *Environ. Pollut.*, *284*, 117105.
[http://dx.doi.org/10.1016/j.envpol.2021.117105] [PMID: 33901981]

Sethiya, N.K., Nahata, A., Mishra, S.H., Dixit, V.K. (2009). An update on Shankhpushpi, a cognition-boosting Ayurvedic medicine. *J. Chin. Integr. Med.*, *7*(11), 1001-1022.
[http://dx.doi.org/10.3736/jcim20091101] [PMID: 19912732]

Sethiya, N.K., Shekh, M.R., Singh, P.K. (2019). Wild banana [*Ensete superbum* (Roxb.) Cheesman.]: Ethnomedicinal, phytochemical and pharmacological overview. *J. Ethnopharmacol.*, *233*, 218-233.
[http://dx.doi.org/10.1016/j.jep.2018.12.048] [PMID: 30686574]

Setyawati, T. (2015). *Narulita Sari, Bahri IP, Raharjo GT. A guidebook of invasive plant species in Indonesia.*. Bogor: Research, Development and Innovation Agency Ministry of Environment and Forestry Republic of Indonesia.

Shah JJ, Jani PM. (1964). Shoot apex of *Euphorbia neriifolia* L. In: *Proc. Natl. Inst. Sci.* India; 30 pp. 81-91.

Shah, K.A., Patel, M.B., Patel, R.J., Parmar, P.K. (2010). *Mangifera indica* (mango). *Pharmacogn. Rev.*, *4*(7), 42-48.
[http://dx.doi.org/10.4103/0973-7847.65325] [PMID: 22228940]

Shah, N.C. (2017). *Phyllanthus emblica* (*Emblica officinalis*): An important medicinal & commercial fruit of India-Part IV. *Life Sci Res.*, *1*(10), 14-23.

Shalgum, A., Govindarajulu, M., Majrashi, M., Ramesh, S., Collier, W.E., Griffin, G., Amin, R., Bradford, C., Moore, T., Dhanasekaran, M. (2019). Neuroprotective effects of *Hibiscus Sabdariffa* against hydrogen peroxide-induced toxicity. *J. Herb. Med., 17-18*, 100253.
[http://dx.doi.org/10.1016/j.hermed.2018.100253]

Shanmugam, S., Manikandan, K., Rajendran, K. (2009). Ethnomedicinal survey of medicinal plants used for the treatment of diabetes and jaundice among the villagers of Sivagangai District, Tamilnadu. *Ethnobot Leaflets., 13*, 189-194.

Shanmugam, S.K., Bama, S., Kiruthiga, N., Kumar, R.S., Sivakumar, T., Dhanabal, P. (2007). Investigation of analgesic activity of leaves part of the *Trianthema portulacastrum* (L) in standard experimental animal models. *Int J Green Pharm., 1*, 39-41.

Shantha, T.R., Patchaimal, P., Reddy, M.P., Kumar, R.K., Tewari, D., Bharti, V., Venkateshwarlu, G., Mangal, A.K., Padhi, M.M., Dhiman, K.S. (2016). Pharmacognostical standardization of Upodika-Basella alba L.: an important ayurvedic antidiabetic plant. *Anc. Sci. Life, 36*(1), 35-41.
[http://dx.doi.org/10.4103/0257-7941.195411] [PMID: 28182032]

Sharma, B.R., (1993). Okra: *Abelmoschus* spp. In *Genetic Improvement of Vegetable Crops* (Eds. Kalloo G, Bergh BO). Pergamon. pp. 751-769.

Sharma, A.M. (2018). A Study of Ethnobotany with Reference to Traditional Knowledge of India. *Int. J. Multidiscip. Res. Sci. Eng. Technol., 1*, 51-57.

Sharma, A.R. (2001). *Sushrut S. Sushrutvimarshini Hindi Commentary along with special Deliberation etc. Part II.* (pp. 311-317). Varanasi: Chaukhambha Surbharati Prakashan.

Sharma, G.K. (2004). Medical ethnobotany in the Svalik range of the Himalayas. *J. Tenn. Acad. Sci., 79*(3), 67-69.

Sharma, M., Neerajarani, G., Mujeeb, C.A., Anu, V., Sravan, B., Kumar, A. (2014). Antioxidant, antifungal and phytochemical analysis of *Bauhinia malabarica*: an *in vitro* study. *Int J Adv Health Sci., 1*(6), 1-3.

Sharma, N., Vijayvergia, R. (2013). A Review on *Digera muricata* (L.) Mart-a great versatile medicinal plant. *Int. J. Pharm. Sci. Rev. Res., 20*(1), 114-119.

Sharma, R.A., Kumari, A. (2014). Phytochemistry, pharmacology and therapeutic application of *Oxalis corniculata* Linn. a review. *Int. J. Pharm. Pharm. Sci., 6*(3), 6-12.

Sharma, V. (2013). Microscopic studies and preliminary pharmacognostical evaluation of *Euphorbia neriifolia* L. leaves. *Indian J. Nat. Prod. Resour., 4*(4), 348-357.

Sharma, Y.M.L. (1980). Bamboo in Asian Pacific Region.*Bamboo research in Asia.* (pp. 99-120). Singapore: World Publication.

Shashi, K., Neetha, C.S. (2019). Evaluation of antidepressant activity of *Tricholepis glaberrima* bark alone and in combination. *Nat J Physiol. Pharm. Pharma., 9*(8), 775-779.

Shawky, E.M., Elgindi, M.R., Ibrahim, H.A., Baky, M.H. (2021). The potential and outgoing trends in traditional, phytochemical, economical, and ethnopharmacological importance of family Onagraceae: A comprehensive review. *J. Ethnopharmacol., 281*, 114450.
[http://dx.doi.org/10.1016/j.jep.2021.114450] [PMID: 34314807]

Sheahan, M.C. (1841). Zygophyllaceae: Zygophyllaceae R. Br., Flind. Voy. Bot. app. 3: 545 (1814).

Balanitaceae Endl.*Flowering Plants· Eudicots: Berberidopsidales, Buxales, Crossosomatales, Fabales pp, Geraniales, Gunnerales, Myrtales pp, Proteales, Saxifragales, Vitales, Zygophyllales, Clusiaceae Alliance, Passifloraceae Alliance, Dilleniaceae, Huaceae, Picramniaceae, Sabiaceae 2007.* (pp. 488-500). Berlin, Heidelberg: Springer Berlin Heidelberg.

Sheeja, E., Joshi, S.B., Jain, D.C. (2010). Bioassay-guided isolation of anti-inflammatory and antinociceptive compound from *Plumbago zeylanica* leaf. *Pharm. Biol., 48*(4), 381-387. [http://dx.doi.org/10.3109/13880200903156424] [PMID: 20645715]

Shen-Miller, J. (2002). Sacred lotus, the long-living fruits of China Antique. *Seed Sci. Res., 12*(3), 131-143. [http://dx.doi.org/10.1079/SSR2002112]

Shenoy, K.R.P., Yoganarasimhan, S.N. (2009). Antibacterial activity of Kutajarista-An ayurvedic preparation. *Indian J. Tradit. Knowl., 8*(3), 362-363.

Shepherd, K.A., Wilson, P.G. (2007). Incorporation of the Australian genera *Halosarcia, Pachycornia, Sclerostegia* and *Tegicornia* into *Tecticornia* (Salicornioideae, Chenopodiaceae). *Aust. Syst. Bot., 20*(4), 319-331. [http://dx.doi.org/10.1071/SB07002]

Shetgiri, N.P., Kokitkar, S.V., Sawant, S.N. (2001). *Radermachera xylocarpa*: the highly efficient source of lapachol and synthesis of its derivatives. *Acta Pol. Pharm., 58*(2), 133-135. [PMID: 11501791]

Shetty, BS, Udupa, SL, Udupa, AL (2008). Biochemical analysis of granulation tissue in steroid and *Centella asiatica* (Linn) treated rats *Pharma online, 2*, 624-632.

S., L.D., Divakar, M. (2014). *Wrightia tinctoria* (Roxb) R.Br. - An updated Review. *Hygeia:journal for drugs and medicines, 6*(1), 95-105. [http://dx.doi.org/10.15254/H.J.D.Med.6.2014.126]

Shi, X.F., Du, D.J., Xie, D.C., Ran, C.Q. (1993). [Studies on the antitumor effect of *Clerodendrum bungei* Steud or C. foetidum Bge]. *Zhongguo Zhongyao Zazhi, 18*(11), 687-690, 704. [PMID: 8003233]

Shi, Z., Chen, Y.L., Chen, Y.S. (2007). Asteraceae (Compositae). In: Wu ZY, Raven PH, Hong DY, editors. Flora of China Volume 20-21 (Asteraceae). Beijing and St. Louis: Science Press and Missouri Botanical Garden.

Shibeshi, W., Makonnen, E., Zerihun, L., Debella, A. (2006). Effect of *Achyranthes aspera* L. on fetal abortion, uterine and pituitary weights, serum lipids and hormones. *Afr. Health Sci., 6*(2), 108-112. [PMID: 16916302]

Shirwaikar, A., Rajendran, K., Barik, R. (2007). Effect of *Garuga pinnata. Pharm. Biol., 45*(3), 205-209. [http://dx.doi.org/10.1080/13880200701213120]

Shisode, K.S., Kareppa, B.M. (2011). *In-vitro* antioxidant activity and phytochemical studies of *Boerhaavia diffusa* linn. roots. Int J Pharma. *Sci. Res., 2*(12), 3171. [http://dx.doi.org/10.13040/IJPSR.0975-8232.2(12).3170-76]

Shivhare, M., Singour, P.K., Chaurasiya, P.K., Pawar, R. (2012). *Trianthema portulacastrum* linn. (bishkhapra). *Pharmacogn. Rev., 6*(12), 132-140. [http://dx.doi.org/10.4103/0973-7847.99947] [PMID: 23055639]

Shivprasad, M., Rane, M., Manik, P. (2016). Traditional uses of some wild edible fruits from Palghar district.

J Nat Prod Plant Resour., 6(6), 8-11.

Shukla, B., Saxena, S., Usmani, S., Kushwaha, P. (2021). Phytochemistry and pharmacological studies of *Plumbago zeylanica* L.: a medicinal plant review. *Clinical Phytoscience, 7*(1), 34.
[http://dx.doi.org/10.1186/s40816-021-00271-7]

Shyma, T.B., Devi Prasad, A.G. (2012). Traditional use of medicinal plants and its status among the tribes in Mananthavady of Wayanad district, Kerala. *World Res J Med Arom Plants, 1*(2), 22-26.

Siddiqua, A., Zahra, M., Begum, K., Jamil, M. (2018). The traditional uses, phytochemistry and pharmacological properties of *Cassia fistula. J. Pharm. Pharmacogn. Res., 2*(1), 15-23.
[http://dx.doi.org/10.26502/jppr.0006]

Siew, Y.Y., Zareisedehizadeh, S., Seetoh, W.G., Neo, S.Y., Tan, C.H., Koh, H.L. (2014). Ethnobotanical survey of usage of fresh medicinal plants in Singapore. *J. Ethnopharmacol., 155*(3), 1450-1466.
[http://dx.doi.org/10.1016/j.jep.2014.07.024] [PMID: 25058874]

Simões, A.R.G., Eserman, L.A., Zuntini, A.R., Chatrou, L.W., Utteridge, T.M.A., Maurin, O., Rokni, S., Roy, S., Forest, F., Baker, W.J., Stefanović, S. (2022). A bird's eye view of the systematics of Convolvulaceae: novel insights from nuclear genomic data. *Front. Plant Sci., 13*, 889988.
[http://dx.doi.org/10.3389/fpls.2022.889988] [PMID: 35909765]

Simon, R.D., Abeliovich, A., Belkin, S. (1994). A novel terrestrial halophilic environment: The phylloplane of *Atriplex halimus*, a salt-excreting plant. *FEMS Microbiol. Ecol., 14*(2), 99-109.
[http://dx.doi.org/10.1111/j.1574-6941.1994.tb00097.x]

Singh, A.K., Raghubanshi, A.S., Singh, J.S. (2002). Medical ethnobotany of the tribals of Sonaghati of Sonbhadra district, Uttar Pradesh, India. *J. Ethnopharmacol., 81*(1), 31-41.
[http://dx.doi.org/10.1016/S0378-8741(02)00028-4] [PMID: 12020925]

Singh, J., Sinha, K., Sharma, A., Mishra, N.P., Khanuja, S.P. (2003). Traditional uses of *Tinospora cordifolia* (Guduchi). *Curr. Res. Med. Aromat. Plants, 25*, 748-751.

Singh, K., Yadava, R.N., Yadav, R. (2023). Antibacterial Compound Isolation and Characterization from the Plant *Cyanotis axillaris* Schult. *Chem. Biodivers., 20*(10), e202301094.
[http://dx.doi.org/10.1002/cbdv.202301094] [PMID: 37690999]

Saiful Yazan, L., Armania, N. (2014). *Dillenia* species: A review of the traditional uses, active constituents and pharmacological properties from pre-clinical studies. *Pharm. Biol., 52*(7), 890-897.
[http://dx.doi.org/10.3109/13880209.2013.872672] [PMID: 24766363]

Singh, M., Kumari, R., Kotecha, M. (2016). *Basella rubra* Linn. - A Rivew. *Int J Ayu Pharm Chem, 5*(1), 206-223.

Singh, P.K., Dwevedi, A.K., Dhakre, G. (2011). Evaluation of antibacterial activities of *Chenopodium album* L. *Int. J. Appl. Biol. Pharm. Technol., 2*(3), 398-401.

Singh, R., Singh, S., Jeyabalan, G., Ali, A. (2012). An overview on traditional medicinal plants as aphrodisiac agent. *J Phacog Phytochem., 1*(4), 43-56.

Singh, S., Jaiswal, S. (2014). Therapeutic properties of *Ficus religiosa. Int J Eng Res Gen Sci., 2*(5), 149-158.

Singh, S., Siew, Y.Y., Yew, H.C., Neo, S.Y., Koh, H.L. (2019). Botany, Phytochemistry and Pharmacological activities of Leea Species.*Medicinal Plants: Chemistry, Pharmacology, and Therapeutic*

Applications. (p. 12). Boca Raton, Florida: CRC Press.
[http://dx.doi.org/10.1201/9780429259968-2]

Singh, S. (2018). Timber yielding plants of district Haridwar and adjacent Siwalik hills. *Ann. Plant Sci., 7*(1), 149-158.

Singh, V., Nimbkar, N. (2007). Safflower (*Carthamus tinctorius* L.). In Genetic resources, chromosome engineering and crop improvement. *Ram J Singh*; 4(6): 167-94.

Sinha, S., Sharma, A., Reddy, P.H., Rathi, B., Prasad, N.V.S.R.K., Vashishtha, A. (2013). Evaluation of phytochemical and pharmacological aspects of *Holarrhena antidysenterica* (Wall.): A comprehensive review. *J. Pharm. Res., 6*(4), 488-492.
[http://dx.doi.org/10.1016/j.jopr.2013.04.004]

Sini, K.R., Sinha, B.N., Rajasekaran, A. (2011). Protective effects of *Capparis zeylanica* Linn. Leaf extract on gastric lesions in experimental animals. *Avicenna J. Med. Biotechnol., 3*(1), 31-35.
[PMID: 23407576]

Sivakumar, D., Chen, L., Sultanbawa, Y. (2018). A comprehensive review on beneficial dietary phytochemicals in common traditional Southern African leafy vegetables. *Food Sci. Nutr., 6*(4), 714-727.
[http://dx.doi.org/10.1002/fsn3.643] [PMID: 29983933]

Sivarajan, VV, Balachandran, I (1994). *Ayurvedic drugs and their plant sources. Oxford and IBH publishing.*

Śmigerska, K. (2016). Research on the Improvement of Growing for Seeds of the Blood Amaranth (*Amaranthus cruentus* L.) of Rawa Variety. Bydgoszcz: Uniwersytet Technologiczno-Przyrodniczy im. *JJ Śniadeckich w Bydgoszczy.*

Sneha, B.D., Raghavendra, N., Harisha, C.R., Acharya, R.N. (2013). Development of random amplified polymorphic DNA markers for authentication of *Rivea hypocrateriformis* (desr.) choisy. *Glob. J. Res. Med. Plants Indig. Med., 2*(5), 348.

Somashekhar, AP, Mishra, SH (2007). Development of random amplified polymorphic DNA markers for authentication of Rivea hypocrateriformis (desr.) choisy. *Global Journal of Research on Medicinal Plants & Indigenous Medicine, 2*(5), 348.

Son, H.L., Yen, P.T.H. (2014). Preliminary phytochemical screening, acute oral toxicity and anticonvulsant activity of the berries of *Solanum nigrum* Linn. *Trop. J. Pharm. Res., 13*(6), 907-912.
[http://dx.doi.org/10.4314/tjpr.v13i6.12]

Song, J., Huang, H., Hao, Y., Song, S., Zhang, Y., Su, W., Liu, H. (2020). Nutritional quality, mineral and antioxidant content in lettuce affected by interaction of light intensity and nutrient solution concentration. *Sci. Rep., 10*(1), 2796.
[http://dx.doi.org/10.1038/s41598-020-59574-3] [PMID: 32071377]

Specht, C.D. (2006). Systematics and evolution of the tropical monocot family Costaceae (Zingiberales): a multiple dataset approach. *Syst. Bot., 31*(1), 89-106.
[http://dx.doi.org/10.1600/036364406775971840]

Sravanakumar, K., Chikatipalli, R., Bonnoth, C.S.K. (2020). Anti-Arthritic activity of *Amaranthus roxburghianus* Nevski hydroalcoholic areal plant extract in acute and chronic models in Albino wistar rats. *Asian J. Pharm. Clin. Res., 13*(6), 112-113.
[http://dx.doi.org/10.22159/ajpcr.2020.v13i6.36994]

Sreedharan, T.P. (2004). *Biological Diversity of Kerala: A survey of Kalliasseri panchayat, Kannur district.*

Kerala Research Programme on Local Level Development. Centre for Development Studies.

Sreena, K.K., Nair, S.S., Mathew, M. (2011). Inventi Rapid. *Ethnopharmacol., 1*(2), 1-4.

Sridhar, K.R., Bhat, R. (2007). Lotus – A potential nutraceutical source. *J. Agric. Technol., 3*(1), 143-155.

Srivastav, S., Singh, P., Mishra, G., Jha, K.K., Khosa, R.L. (2011). *Achyranthes aspera*-An important medicinal plant: A review. *J Nat Prod Plant Resour., 1*(1), 1-4.

Srivastava, P.K. (2014). *Achyranthes aspera*: A potent immunostimulating plant for traditional medicine. *Int. J. Pharm. Sci. Res., 5*(5), 1601-11.
[http://dx.doi.org/10.13040/IJPSR.0975-8232]

Srivastava, R. (2014). A review on phytochemical, pharmacological, and pharmacognostical profile of *Wrightia tinctoria*: Adulterant of kurchi. *Pharmacogn. Rev., 8*(15), 36-44.
[http://dx.doi.org/10.4103/0973-7847.125528] [PMID: 24600194]

Stanely Mainzen Prince, P., Priscilla, H., Devika, P.T. (2009). Gallic acid prevents lysosomal damage in isoproterenol induced cardiotoxicity in Wistar rats. *Eur. J. Pharmacol., 615*(1-3), 139-143.
[http://dx.doi.org/10.1016/j.ejphar.2009.05.003] [PMID: 19450577]

Srivastava, S, Singh, P, Mishra, G, Jha, KK, Khosa, RL (2011). *Costus speciosus* (Keukand): a review. Der Pharmacia Sinica; 2(1): 118–128.

Ssenyonga, M., Brehony, E. (1993). Herbal medicine-its use in treating some symptoms of AIDS. *Int. Conf. AIDS., 9*, 75-75.

Stagos, D., Amoutzias, G.D., Matakos, A., Spyrou, A., Tsatsakis, A.M., Kouretas, D. (2012). Chemoprevention of liver cancer by plant polyphenols. *Food Chem. Toxicol., 50*(6), 2155-2170.
[http://dx.doi.org/10.1016/j.fct.2012.04.002] [PMID: 22521445]

Stanely Mainzen Prince, P., Menon, V.P. (2001). Antioxidant action of *Tinospora cordifolia* root extract in alloxan diabetic rats. *Phytother. Res., 15*(3), 213-218.
[http://dx.doi.org/10.1002/ptr.707] [PMID: 11351355]

Sudjaroen, Y., Haubner, R., Würtele, G., Hull, W.E., Erben, G., Spiegelhalder, B., Changbumrung, S., Bartsch, H., Owen, R.W. (2005). Isolation and structure elucidation of phenolic antioxidants from Tamarind (*Tamarindus indica* L.) seeds and pericarp. *Food Chem. Toxicol., 43*(11), 1673-1682.
[http://dx.doi.org/10.1016/j.fct.2005.05.013] [PMID: 16000233]

Suguna, L., Sivakumar, P., Chandrakasan, G. (1996). Effects of *Centella asiatica* extract on dermal wound healing in rats. *Indian J. Exp. Biol., 34*(12), 1208-1211.
[PMID: 9246912]

Sujitha, R., Bhimba, B.V., Sindhu, M.S., Arumugham, P. (2013). Phytochemical Evaluation and Antioxidant Activity of *Nelumbo nucifera, Acorus calamus* and *Piper longum. Int J Pharma Chem Sci., 2*(3), 1573-1578.

Sun, J., Blaskovich, M.A., Jove, R., Livingston, S.K., Coppola, D., Sebti, S.M. (2005). Cucurbitacin Q: a selective STAT3 activation inhibitor with potent antitumor activity. *Oncogene, 24*(20), 3236-3245.
[http://dx.doi.org/10.1038/sj.onc.1208470] [PMID: 15735720]

Suresh, K., Deepa, P., Harisaranraj, R., Vaira Achudhan, V. (2008). Antimicrobial and phytochemical investigation of the leaves of *Carica papaya* L., *Cynodon dactylon* (L.) Pers., *Euphorbia hirta* L., *Melia azedarach* L. and *Psidium guajava* L. *Ethnobot Leaflets., 12*, 1184-1189.

Sureshkumar, S.V., Mishra, S.H. (2007). Hepatoprotective activity of extracts from *Pergularia daemia* (Forsk) against carbon tetrachloride-induced toxicity in rats. *Pharmacogn. Mag., 3*, 187-191.

Sutar, N.G., Sutar, U.N., Sharma, Y.P., Shaikh, I.K., Kshirsagar, S.S. (2008). Phytochemical investigation and pharmacological screening of leaves of *Achyranthes aspera* Linn. as analgesic and antipyretic. *Biosci. Biotechnol. Res. Asia, 5*(2), 841-844.

Talal Al-Malki, A, Abbas Ahmed Sindi, H, Hussain Al-Qahtani, M. (2021). Regulation of CD36 gene expression by *Hibiscus sabdariffa* tea extracts to affect the atherosclerosis biomarkers in Saudi women. *J. Food Nutr. Res.*; 9:4 5–49.
[http://dx.doi.org/10.12691/jfnr-9-1-7]

Tambekar, D.H., Dahikar, S.B. (2010). Exploring antibacterial potential of some ayurvedic preparations to control bacterial enteric infections. *J. Chem. Pharm. Res., 2*(5), 494-501.

Tan, C.X., Tan, S.T., Tan, S.S. (2020). An overview of papaya seed oil extraction methods. *Int. J. Food Sci. Technol., 55*(4), 1506-1514.
[http://dx.doi.org/10.1111/ijfs.14431]

Tang, Y., Xin, H., Guo, M. (2016). Review on research of the phytochemistry and pharmacological activities of *Celosia argentea. Rev. Bras. Farmacogn., 26*(6), 787-796.
[http://dx.doi.org/10.1016/j.bjp.2016.06.001]

Tasduq, S.A., Kaisar, P., Gupta, D.K., Kapahi, B.K., Jyotsna, S., Johri, R.K., Johri, R.K. (2005). Protective effect of a 50% hydroalcoholic fruit extract ofEmblica officinalis against anti-tuberculosis drugs induced liver toxicity. *Phytother. Res., 19*(3), 193-197.
[http://dx.doi.org/10.1002/ptr.1631] [PMID: 15934014]

Telrandhe, U.B., Gunde, M.C. (2022). Phytochemistry, pharmacology and multifarious activity of *Cassia tora* L.: A comprehensive review. *Ann. Phytomed., 11*(2), 231-239.
[http://dx.doi.org/10.54085/ap.2022.11.2.25]

Tenni, R., Zanaboni, G., De Agostini, M.P., Rossi, A., Bendotti, C., Cetta, G. (1988). Effect of the triterpenoid fraction of *Centella asiatica* on macromolecules of the connective matrix in human skin fibroblast cultures. *Ital. J. Biochem., 37*(2), 69-77.
[PMID: 3042688]

Tewodros, M. (2013). Genetic diversity of Taro (*Colocasia esculenta* (L.) Schott) genotypes in Ethiopia based on agronomic traits. *Time J. Agric. Vet. Sci., 1*, 23-30.

Thakur, M., Bhargava, S., Dixit, V.K. (2007). Immunomodulatory Activity of *Chlorophytum borivilianum* Sant. F. *Evid. Based Complement. Alternat. Med., 4*(4), 419-423.
[http://dx.doi.org/10.1093/ecam/nel094] [PMID: 18227908]

Thakur, M., Dixit, V.K. (2006). Effect of *Chlorophytum borivilianum* on androgenic and sexual behavior of male rats. *Indian Drugs., 43*, 300-306.

Thamizhselvam, N, Gk, S. (2020). Medicinal plants in Rasayana drugs, their active ingredients and reported biological activities: an overview. *J. Nat. Ayurved. Med.*, 4(1).

The Wealth of India–A Dictionary Indian Raw Materials and Industrial Products: Raw Materials Series, Revised edn, Vol. 3 Ca-Ci, Publications & Information Directorate, CSIR, New Delhi, 1992; pp. 276-293.

Thenmozhi, K., Manian, S., Paulsamy, S. (2013). Antioxidant and free radical scavenging potential of leaf and stem bark extracts of *Bauhinia malabarica* Roxb Int. *J. Pharm. Sci.,* *5*(1), 306-311.

Thet, K.K., Zar, K., Lae, W., Phyo, H.W., Khin, A.N. (2020). Nutritional values and antioxidant activity analysis of *Merremia emarginata* (Burm.F). 3rd Myanmar Korea Conf. *R J., 3*, 1549-1555.

Thissera, B., Visvanathan, R., Khanfar, M.A., Qader, M.M., Hassan, M.H.A., Hassan, H.M., Bawazeer, M., Behery, F.A., Yaseen, M., Liyanage, R., Abdelmohsen, U.R., Rateb, M.E. (2020). *Sesbania grandiflora* L. Poir leaves: A dietary supplement to alleviate type 2 diabetes through metabolic enzymes inhibition. *S. Afr. J. Bot., 130*, 282-299.
[http://dx.doi.org/10.1016/j.sajb.2020.01.011]

Thorat, B.R. (2018). Review on *Celosia argentea* L. *Plant. Res J Phrama Phytochem., 10*(1), 109-119.
[http://dx.doi.org/10.5958/0975-4385.2018.00017.1]

Tirimana, A.S.L. (1987). *Medicinal plants of suriname. Uses and chemical constituents.* (p. 92). Suriname: Chemical Laboratory, Ministry of Agriculture, Animal Husbandry and Fisheries.

Tiwari, SK (2021). Diversity and Socioeconomic Importance of Different Medicinal Plants in Korba Chhattisgarh India. *Frontiers in Science and Technology in India,* 50-5.

Townsend, C.C. Amaranthaceae. In: Flora of Tropical East Africa, [ed. by Polhill RM]. Rotterdam, Netherlands: A.A. Balkema 1985; 1-2, 20-24, 35-36.

Trichopoulos, D., Willett, W.C. (1996). Introduction: Nutrition and cancer. *Cancer Causes Control, 7*(1), 3-4.
[http://dx.doi.org/10.1007/BF00115633] [PMID: 8998310]

Tripathi, K., Gore, P.G., Pandey, A., Nayar, E.R., Gayacharan, C., Pamarthi, R.K., Bhardwaj, R., Kumar, A. (2021). Morphological and nutritional assessment of *Vigna vexillata* (L.) A. Rich.: a potential tuberous legume of India. *Genet. Resour. Crop Evol., 68*(1), 397-408.
[http://dx.doi.org/10.1007/s10722-020-01023-1]

Tundis, R., Rashed, K., Said, A., Menichini, F., Loizzo, M.R. (2014). *In vitro* cancer cell growth inhibition and antioxidant activity of *Bombax ceiba* (Bombacaceae) flower extracts. *Nat. Prod. Commun., 9*(5), 1934578X1400900.
[http://dx.doi.org/10.1177/1934578X1400900527] [PMID: 25026723]

Tyler, V. (1994). *Herbs of choice. The therapeutic use of phytomedicinals.* (3rd ed.). New York: Haworth Press.

Udayan, P.S., Harinarayanan, M.K., Tushar, K.V., Indira, B. (2008). Some common plants used by Kurichiar tribes of Tirunelli forest, Wayanad district, Kerala in medicine and other traditional uses. *Indian J. Tradit. Knowl., 7*, 250-255.

Uddin, M.M., Chen, Z., Huang, L. (2020). Cadmium accumulation, subcellular distribution and chemical fractionation in hydroponically grown *Sesuvium portulacastrum* [Aizoaceae]. *PLoS One, 15*(12), e0244085.
[http://dx.doi.org/10.1371/journal.pone.0244085] [PMID: 33370774]

Udupa AL, Rathnakar UP, Udupa S. (2007). Ind. drugs., 44(6): 466- 469.

Umar, KJ, Hassan, LG, Dangoggo, SM, Maigandi, SA, Sani, NA (2011). Nutritional and anti-nutritional profile of spiny amaranth (*Amaranthus viridis* Linn). Studia Universitatis Vasile Goldis Seria Stiintele Vietii (Life Sciences Series). 21(4).

Umdale, S.D., Chavan, J.J., Ahire, M.L., Kshirsagar, P.R., Gaikwad, N.B., Bhat, K.V. (2018). *Vigna*

khandalensis (Santapau) Raghavan et Wadhwa: a promising underutilized, wild, endemic legume of the Northern Western Ghats, India. *Genet. Resour. Crop Evol., 65*(6), 1795-1807.
[http://dx.doi.org/10.1007/s10722-018-0657-y]

Upadhyay, A., Kumar, K., Kumar, A., Mishra, H. (2010). *Tinospora cordifolia* (Willd.) Hook. f. and Thoms. (Guduchi) - validation of the Ayurvedic pharmacology through experimental and clinical studies. *Int. J. Ayurveda Res., 1*(2), 112-121.
[http://dx.doi.org/10.4103/0974-7788.64405] [PMID: 20814526]

Upadhyay, B., Parveen, A.K., Dhaker, A.K., Kumar, A. (2010). Ethnomedicinal and ethnopharmaco-statistical studies of Eastern Rajasthan, India. *J. Ethnopharmacol., 129*(1), 64-86.
[http://dx.doi.org/10.1016/j.jep.2010.02.026] [PMID: 20214972]

US Forest Service, (2011). *Mucuna pruriens* (L.) DC. Pacific Island Ecosystems at Risk (PIER). USDA-ARS, 2016.

Usha, K., Kasturi, G.M., Hemalatha, P. (2007). Hepatoprotective effect of *Hygrophila spinosa* and *Cassia occidentalis* on carbon tetrachloride induced liver damage in experimental rats. *Indian J. Clin. Biochem., 22*(2), 132-135.
[http://dx.doi.org/10.1007/BF02913331] [PMID: 23105700]

Vadivel, V., Janardhanan, K. (2004). The nutritional and antinutritional attributes of sword bean [*Canavalia gladiata* (Jacq.) DC.]: an under-utilized tribal pulse from south India. *Int. J. Food Sci. Technol., 39*(9), 917-926.
[http://dx.doi.org/10.1111/j.1365-2621.2004.00851.x]

Vijaya Padma, V., Sowmya, P., Arun Felix, T., Baskaran, R., Poornima, P. (2011). Protective effect of gallic acid against lindane induced toxicity in experimental rats. *Food Chem. Toxicol., 49*(4), 991-998.
[http://dx.doi.org/10.1016/j.fct.2011.01.005] [PMID: 21219962]

Van Damme, E., Goossens, K., Smeets, K., Van Leuven, F., Verhaert, P., Peumans, W.J. (1995). The major tuber storage protein of araceae species is a lectin. Characterization and molecular cloning of the lectin from *Arum maculatum* L. *Plant Physiol., 107*(4), 1147-1158.
[http://dx.doi.org/10.1104/pp.107.4.1147] [PMID: 7770523]

van Jaarsveld, P., Faber, M., van Heerden, I., Wenhold, F., Jansen van Rensburg, W., van Averbeke, W. (2014). Nutrient content of eight African leafy vegetables and their potential contribution to dietary reference intakes. *J. Food Compos. Anal., 33*(1), 77-84.
[http://dx.doi.org/10.1016/j.jfca.2013.11.003]

Vasundharan, S.K., Jaishanker, R.N., Annamalai, A., Sooraj, N.P. (2015). Ethnobotany and distribution status of *Ensete superbum* (roxb.) Cheesman in India: A geo-spatial review. *J. Ayurvedic Herb. Med., 1*(2), 54-58.
[http://dx.doi.org/10.31254/jahm.2015.1208]

Vedavathy, S., Rao, D.N. (1995). Herbal folk medicine of Tirumala and Tirupati Region of Chittoor. District, Andhra Pradesh. *Fitoterapia, 66*, 167-171.

Velusamy, P., Kumar, G.V., Jeyanthi, V., Das, J., Pachaiappan, R. (2016). Bio-inspired green nanoparticles: synthesis, mechanism, and antibacterial application. *Toxicol. Res., 32*(2), 95-102.
[http://dx.doi.org/10.5487/TR.2016.32.2.095] [PMID: 27123159]

Venkatachalapathi, A (2015). Thekkan Sangeeth, Paulsamy S. Ethnobotanical informations on the species of selected areas in Nilgiri Biosphere Reserve, the Western Ghats, *India. J. Res. Biol.* 5(A): 43–57.

Venkateshwarlu, M., Nagaraju, M., Odelu, G., Srilatha, T., Ugandhar, T. (2017). Studies on phytochemical analysis and biological activities in *Momordica dioica* Roxb through Fruit. *Pharma Innov. J, 6*(12), 437-440.

Venugopalan, S.N., Venkatasubramanian, P. (2017). Understanding the concepts rasayana in Ayurveda biology. J Nat. *Ayurved. Med., 1*(2), 000112.

Verma, A., Bala, N., Khangembam, N. (2017). Effect of cooking methods on anti-nutrient content of Loni (*Portulaca quadrifida* L.) and their products development. *J. Dairy. Foods Home Sci., 36*(4), 345-348. [http://dx.doi.org/10.18805/ajdfr.DR-1229]

Vijaya Kumar, S., Sankar, P., Varatharajan, R. (2009). Anti-inflammatory activity of roots of *Achyranthes aspera*. *Pharm. Biol., 47*(10), 973-975. [http://dx.doi.org/10.1080/13880200902967979]

vijayakumar, S., Dhanapal, R., Sarathchandran, I., Kumar, A.S., Ratna, J.V. (2012). Evaluation of antioxidant activity of Ammania baccifera (L.) Whole plant extract in rats. *Asian Pac. J. Trop. Biomed., 2*(2), S753-S756. [http://dx.doi.org/10.1016/S2221-1691(12)60309-8]

Vijetha, P.K., Hedge, K. (2018). Shab araya AR. Pharmacological Review on *Spondias pinnata*: the Indian Hog Plum. Inter. *J Pharm. Chem. Res, 4*(2), 141-145.

Vincent, S., Vijay, A.R., Jeevanantham, P., Ragavan, S. (2012). *In-vitro* and *in-vivo* anti-asthmatic activity of *Clerodendrum phlomidis* Linn. In guinea pigs. *Int J Res Rev Pharm Appl Sci., 2*, 15-28.

Vishnukanta, S., Rana, A.C. (2011). *Plumbago zeylanica*: a phytopharmacological review. *Int. J. Pharm. Sci. Res., 2011*, 247-255. [http://dx.doi.org/10.13040/IJPSR.0975-8232]

Vitalone, A., Bordi, F., Baldazzi, C., Mazzanti, G., Saso, L., Tita, B. (2001). Anti-proliferative effect on a prostatic epithelial cell line (PZ-HPV-7) by *Epilobium angustifolium* L. *Farmaco, 56*(5-7), 483-489. [http://dx.doi.org/10.1016/S0014-827X(01)01067-9] [PMID: 11482783]

Vogl, S., Picker, P., Mihaly-Bison, J., Fakhrudin, N., Atanasov, A.G., Heiss, E.H., Wawrosch, C., Reznicek, G., Dirsch, V.M., Saukel, J., Kopp, B. (2013). Ethnopharmacological *in vitro* studies on Austria's folk medicine—An unexplored lore *in vitro* anti-inflammatory activities of 71 Austrian traditional herbal drugs. *J. Ethnopharmacol., 149*(3), 750-771. [http://dx.doi.org/10.1016/j.jep.2013.06.007] [PMID: 23770053]

Wadhawa, G.C., Shivankar, V.S., Munshi, A., Mirgane, N.A. (2020). Antioxidant capacity of *Pimpinella wallichiana*. Int J. Bot. Stud. (Taipei, Taiwan), *5*(6), 396-399.

Wahi, A.K., Ravi, J., Hemalatha, S., Singh, P.N. (2002). Antidiabetic activity of *Daemia extensa*. J. Nat. Remedie., *2*, 80-83.

Wan Ibrahim, W.M.H., Mohamad Amini, M.H., Sulaiman, N.S., Wan Abdul Kadir, W.R. (2021). Evaluation of alkaline-based activated carbon from *Leucaena Leucocephala* produced at different activation temperatures for cadmium adsorption. *Appl. Water Sci., 11*(1), 1-3. [http://dx.doi.org/10.1007/s13201-020-01330-z]

Wang, G., Li, Y. (1985). Clinical application of safflower (*Carthamus tinctorius*). *Zhejiang Tradit Chin Med Sci J., 1*, 42-43.

Wang, G.K., Lin, B.B., Qin, M.J. (2014). [Study on chemical constituents from leaf of *Bombax ceiba* (II)]. *Zhong Yao Cai, 37*(2), 240-242.

[PMID: 25095343]

Wang, J., Li, P., Li, B., Guo, Z., Kennelly, E.J., Long, C. (2014). Bioactivities of compounds from *Elephantopus scaber*, an ethnomedicinal plant from Southwest China. *Evid. Based Complement. Alternat. Med., 2014*(1), 569594.
[http://dx.doi.org/10.1155/2014/569594] [PMID: 24963325]

Wang, J.L., Meng, X., Lu, R., Wu, C., Luo, Y-T., Yan, X., Li, X-J., Kong, X-H., Nie, G-X. (2015). Effects of Rehmannia glutinosa on growth performance, immunological parameters and disease resistance to *Aeromonas hydrophila* in common carp (*Cyprinus carpio* L.). *Aquaculture, 435*, 293-300.
[http://dx.doi.org/10.1016/j.aquaculture.2014.10.004]

Wang, Q., Zhang, X. (2005). *Colored Illustration of Lotus Cultivars in China..* Beijing, China: China Forestry Publishing House.

Wang, Z., Li, J., Ji, Y., An, P., Zhang, S., Li, Z. (2013). Traditional herbal medicine: a review of potential of inhibitory hepatocellular carcinoma in basic research and clinical trial. *Evid. Based Complement. Alternat. Med., 2013*, 1-7.
[http://dx.doi.org/10.1155/2013/268963] [PMID: 23956767]

Warrier, P.K. (1996). Indian medicinal plants, compendium of 500 species. *Orient Longman Ltd., 5*, 173.

Warries, P.K., Nambiar, V.P., Raman, K.C. (1995). Indian medicinal plants: A compendium of 500 species. *Orient Longman, 3*, 165.

Waterhouse, B., Mitchell, A. (1998). *Northern Australia Quarantine Strategy Weeds Target List.* (p. 29). Canberra: AQIS Miscellaneous Publication.

Watt G. (1972). A Dictionary of the Economic Products of India, reprinted edition, volume-I, Periodical Expert, Delhi; 383-391.

Wazi, S.M., Saima, S., Dasti, A.A., Subhan, S. (2007). Ethnobotanical importance of Salt range species of District Karak, Pakistan. Pakistan. *J. Plant Sci., 13*, 29-31.

Wealth of India Council of Scientific and Industrial Research. (1953). *Publication and Information Directorate.* (pp. 54-55). New Delhi.

Wen J. Vitaceae: Vitaceae Juss., Gen. Pl.: 267 (1789), nom. cons. InFlowering Plants• Eudicots: Berberidopsidales, Buxales, Crossosomatales, Fabales pp, Geraniales, Gunnerales, Myrtales pp, Proteales, Saxifragales, Vitales, Zygophyllales, Clusiaceae Alliance, Passifloraceae Alliance, Dilleniaceae, Huaceae, Picramniaceae, Sabiaceae 2007 (pp. 467-479). Berlin, Heidelberg: Springer Berlin Heidelberg.

Wesley, J.J., Christina, A.J., Chidambaranathan, N. (2008). Effect of alcoholic extract of *Tinospora Cordifolia* on acute and subacute Inflammation. *Pharmacologyonlin, 3*, 683-687.

Wheeler, R.A., Chaney, W.R., Johnson, K.D., Butler, L.G. (1996). Leucaena forage analysis using near infrared reflectance spectroscopy. *Anim. Feed Sci. Technol., 64*(1), 1-9.
[http://dx.doi.org/10.1016/S0377-8401(96)01047-4]

White Wild Musk Mallow. Flowers of India. Retrieved 29 July 2010.

Wiart C. Medicinal plants of the Asia-Pacific: drugs for the future? World Scientific; 2006 Jan 11.
[http://dx.doi.org/10.1142/5834]

Wiese, J.L., Meadow, J.F., Lapp, J.A. (2012). Seed weights for northern Rocky Mountain native plants with

an emphasis on Glacier National Park. *Native Plants J., 13*(1), 39-50.
[http://dx.doi.org/10.3368/npj.13.1.39]

Williamson, E.M. (2002). *Major herbs of Ayurveda.*. London, UK: Churchill Livingstone.

Williamson, K.M., Nadkarni, A.K. (1954). *The Indian Materia Medica.* Bombay Popular Prakashan.

Winter RA. (2009). Consumer's Dictionary of Cosmetic Ingredients. 7th Edition: Complete Information About the Harmful and Desirable Ingredients Found in Cosmetics and Cosmeceuticals. Three Rivers Press; New York, NY, USA.

Wright, K.H., Pike, O.A., Fairbanks, D.J., Huber, C.S. (2002). Composition of *Atriplex hortensis*, sweet and bitter *Chenopodium quinoa* seeds. *J. Food Sci., 67*(4), 1383-1385.
[http://dx.doi.org/10.1111/j.1365-2621.2002.tb10294.x]

Xu, Z., Chang, L. (2017). Dioscoreaceae. *Identification and Control of Common Weeds, 3*, 895-903.
[http://dx.doi.org/10.1007/978-981-10-5403-7_36]

Xu, Z., Chang, L. (2017). Rubiaceae. *Identification and Control of Common Weeds.* (Vol. 3). Singapore: Springer.
[http://dx.doi.org/10.1007/978-981-10-5403-7_16]

Xu, Z., Deng, M., Xu, Z., Deng, M. (2017). Lythraceae. *Identification and Control of Common Weeds, 2*, 765-784.

Xu, Z., Deng, M. (2017). Nyctaginaceae. *Identification and Control of Common Weeds, 2*, 303-308.

Xu, Z., Deng, M. (2017). Portulacaceae. *Identification and Control of Common Weeds, 2*, 319-323.

Xue, J.H., Dong, W.P., Cheng, T., Zhou, S.L. (2012). Nelumbonaceae: Systematic position and species diversification revealed by the complete chloroplast genome. *J. Syst. Evol., 50*(6), 477-487.
[http://dx.doi.org/10.1111/j.1759-6831.2012.00224.x]

Yadav, R., Rai, R., Yadav, A., Pahuja, M., Solanki, S., Yadav, H. (2016). Evaluation of antibacterial activity of *Achyranthes aspera* extract against *Streptococcus mutans*: An *in vitro* study. *J. Adv. Pharm. Technol. Res., 7*(4), 149-152.
[http://dx.doi.org/10.4103/2231-4040.191426] [PMID: 27833895]

Yadava, R.N., Prashant, B. (2009). New antifungal constituent from the *Tricholepsis glaberrima* DC. *Asian J. Chem., 21*(9), 6683-6688.

Yamaki, J., Venkata, K.C.N., Mandal, A., Bhattacharyya, P., Bishayee, A. (2016). Health-promoting and disease-preventive potential of *Trianthema portulacastrum* Linn. (Gadabani)—An Indian medicinal and dietary plant. *J. Integr. Med., 14*(2), 84-99.
[http://dx.doi.org/10.1016/S2095-4964(16)60247-9] [PMID: 26988430]

Yamauchi, D., Minamikawa, T. (1987). Synthesis of canavalin and concanavalin A in maturing *Canavalia gladiata* seeds. *Plant Cell Physiol., 28*(3), 421-430.

Yen, K. (1992). *The Illustrated Chinese Materia Medica.* (p. 84). Taipei: SMC Publishing.

Yeshwante, S.B., Juvekar, A.R., Nagmoti, D.M., Wankhede, S.S., Shah, A.S., Pimprikar, R.B., Saindane, D.S. (2009). Anti-inflammatory activity of methanolic extracts of *Dillenia indica* L. leaves. *J. Young Pharm., 1*(1), 63-66.

[http://dx.doi.org/10.4103/0975-1483.51885]

Yesodharan, K., Sujana, K.A. (2007). Status of ethnomedicinal plants in the Parambikulam wildlife sanctuary, Kerala, South India. *Ann. For. (Dehra Dun), 15*(2), 322-334.

Yoganarasimhan, S.N. (2000). *Medicinal plants of Ind ia.* (p. 1). Bangalore: Interline Publishing Pvt. Ltd..

Yousif, M.A.I., Wang, Y.R., Dali, C. (2020). Seed dormancy overcoming and seed coat structure change in *Leucaena leucocephala* and *Acacia nilotica*. *Forest Sci. Technol., 16*(1), 18-25.
[http://dx.doi.org/10.1080/21580103.2019.1700832]

Yuan, L.P., Chen, F.H., Ling, L., Dou, P.F., Bo, H., Zhong, M.M., Xia, L.J. (2008). Protective effects of total flavonoids of *Bidens pilosa* L. (TFB) on animal liver injury and liver fibrosis. *J. Ethnopharmacol., 116*(3), 539-546.
[http://dx.doi.org/10.1016/j.jep.2008.01.010] [PMID: 18313245]

Yusuff, A.S., Lala, M.A., Popoola, L.T., Adesina, O.A. (2019). Optimization of oil extraction from *Leucaena leucocephala* seed as an alternative low-grade feedstock for biodiesel production. *SN Applied Sciences, 1*(4), 357.
[http://dx.doi.org/10.1007/s42452-019-0364-0]

Yeap Foo, L., Wong, H. (1992). Phyllanthusiin D, an unusual hydrolysable tannin from *Phyllanthus amarus*. *Phytochemistry, 31*(2), 711-713.
[http://dx.doi.org/10.1016/0031-9422(92)90071-W]

Zahan, R., Nahar, L., Haque, M., Nesa, M.L., Alam, Z. (2013). Antioxidant and Antidiuretic Activities of *Alangium salvifolium* and *Bombax ceiba*. Dhaka Univ. *J. Pharm. Sci., 12*(2), 159-163.

Zahara, K., Bibi, Y., Tabassum, S., Mudrikah, , Bashir, T., Haider, S., Araa, A., Ajmal, M. (2015). A review on pharmacological properties of *Bidens biternata*: A potential nutraceutical. *Asian Pac. J. Trop. Dis., 5*(8), 595-599.
[http://dx.doi.org/10.1016/S2222-1808(15)60894-5]

Zahin, M., Ahmad, I., Aqil, F. (2010). Antioxidant and antimutagenic activity of *Carum copticum* fruit extracts. *Toxicol. In Vitro, 24*(4), 1243-1249.
[http://dx.doi.org/10.1016/j.tiv.2010.02.004] [PMID: 20149861]

Zaier, H., Ghnaya, T., Lakhdar, A., Baioui, R., Ghabriche, R., Mnasri, M., Sghair, S., Lutts, S., Abdelly, C. (2010). Comparative study of Pb-phytoextraction potential in *Sesuvium portulacastrum* and *Brassica juncea*: Tolerance and accumulation. *J. Hazard. Mater., 183*(1-3), 609-615.
[http://dx.doi.org/10.1016/j.jhazmat.2010.07.068] [PMID: 20708335]

Zaman, S. (2016). *In vitro* evaluation of antimicrobial, antioxidant and cytotoxic activities of *Garuga pinnata* leaves. *Indo American J Pharma Sci., 3*(3), 294-299.

Zaoui, A., Cherrah, Y., Mahassini, N., Alaoui, K., Amarouch, H., Hassar, M. (2002). Acute and chronic toxicity of *Nigella sativa* fixed oil. *Phytomedicine, 9*(1), 69-74.
[http://dx.doi.org/10.1078/0944-7113-00084]

Zarena, A.S., Gopal, S., Vineeth, R. (2014). Antioxidant, antibacterial, and cytoprotective activity of agathi leaf protein. *J. Anal. Methods Chem., 2014*, 1-8.
[http://dx.doi.org/10.1155/2014/989543] [PMID: 24616824]

Zargari, A. (1996). *Herb.* (6th ed., p. 2). Tehran, Iran: Tehran University.

Zarshenas, M.M., Moein, M., Samani, S.M., Petramfar, P. (2013). An overview on ajwain (*Trachyspermum ammi*) pharmacological effects; modern and traditional. *J. Nat. Rem., 14*(1), 98-105.

Zayed, M.Z., Sallam, S.M., Shetta, N.D. (2018). Review article on *Leucaena leucocephala* as one of the miracle timber trees. *Int J Pharma Pharma. Sci., 10*(1), 1-7.

Zhang, Y.J., Abe, T., Tanaka, T., Yang, C.R., Kouno, I. (2001). Phyllanemblinins A-F, new ellagitannins from *Phyllanthus emblica. J. Nat. Prod., 64*(12), 1527-1532.
[http://dx.doi.org/10.1021/np010370g] [PMID: 11754604]

Zhou, F.R., Zhao, M.B., Tu, P.F. (2009). Simultaneous determination of four nucleosides in *Carthamus tinctorius* L. and Safflower injection using high performance liquid chromatography. *J Chin Pharmaceut Sci (Chin), 18*, 326-330.

SUBJECT INDEX

A

Abelmoschus ficulneus 51
Acid 5, 19, 22, 28, 31, 44, 47, 50, 53, 56, 59, 61, 64, 70, 80, 88, 94, 96, 97, 104, 106, 112, 119, 130, 132, 134, 137, 142, 149, 156, 160, 163, 167, 169, 176
  acetic 53
  alpha-linolenic (ALA) 134, 137
  amaretto 53
  amino 22, 44, 47, 80, 142, 160
  ascorbic 22
  betulinic 31, 50, 61
  caffeic 130
  clerodermic 163
  chlorogenic 59, 61
  chrysophanic 28
  cis-coumaric 19
  citric 64
  ellagic 50, 99, 112, 176
  fatty 22, 47, 59, 132, 134, 149
  ferulic 61, 130, 134
  gallic 5, 112, 156, 167, 169
  gamma-linolenic 97
  gentisic 156
  gluconic 88
  hibiscus 64
  linoleic 22
  malic 64, 88
  mollic 167
  nicotinic 31
  oleanolic 28, 61, 96, 119, 163
  oleic 94, 160
  oxalic 19, 104, 106
  palmitic 53, 160
  protocatechuic 56, 59, 156
  serratagenic 163
  syringic 169
  tartaric acid 64, 88
  tinosporic 70
  trans-coumaric 19
  triterpenoidsbetulinic 28, 96

  ursolic 96, 167
Activity 6, 9, 11, 50, 56, 57, 92, 93, 109, 112, 123, 124, 153
  anti-diabetic 93
  anti-inflammatory 50, 92, 153
  antimicrobial 6, 9, 11, 56, 109, 124
  antioxidant 6, 11, 50, 57, 92, 112, 123
  diuretic 92
  hepatoprotective 92
Agroforestry systems 20, 28, 29, 32
Alasunda 42
Alkaloids 10, 13, 19, 22, 25, 28, 56, 67, 68, 77, 80, 90, 92, 96, 115, 142, 145, 149, 167, 169, 176
  bombaxine 56
  brahmine 115
  catharticin 28
  flavones 149
  harmine 176
  mimosine 22
  schreberine 96
Alley cropping 20
Ammannia baccifera 48
Animal feed 135, 137
Anonaine 88
Anthocyanins 52, 64, 65, 106
Anthraquinone 16, 28, 130
  chrysophanol 130
  dimethoxy-2-methyle 16
  sennosides 28
Antioxidants 17, 45, 47, 59, 64, 65, 135, 153, 154, 164, 165, 173, 177
Apigenin 59, 156, 163, 176
Aquatic perennial plant 87
Arabinogalactan 70
Arabinoside 167
Aromatherapy 54
Arthritis 92, 96, 120, 149, 153
Asthma 11, 62, 78, 103, 119, 164

Ganesh Chandrakant Nikalje, Apurva Chonde, Sudhakar Srivastava & Penna Suprasanna

www.ingramcontent.com/pod-product-compliance
Lightning Source LLC
Chambersburg PA
CBHW050823220326
41598CB00006B/301